Fundamental Research and Application of Droplet Dynamics

Edited by Hongliang Luo

Published in London, United Kingdom

Fundamental Research and Application of Droplet Dynamics
http://dx.doi.org/10.5772/intechopen.97996
Edited by Hongliang Luo

Contributors
Muhammad M. Iqbal, Mark M. Turner, PengBo Dong, Raza Gulfam, Hongliang Luo, Feixiang Chang, Anu Osta

First published in London, United Kingdom, 2022 by IntechOpen
IntechOpen is the global imprint of INTECHOPEN LIMITED, registered in England and Wales, registration number: 11086078, 5 Princes Gate Court, London, SW7 2QJ, United Kingdom

British Library Cataloguing-in-Publication Data
A catalogue record for this book is available from the British Library

Additional hard and PDF copies can be obtained from orders@intechopen.com

Fundamental Research and Application of Droplet Dynamics
Edited by Hongliang Luo
p. cm.
Print ISBN 978-1-80355-960-5
Online ISBN 978-1-80355-961-2
eBook (PDF) ISBN 978-1-80355-962-9

Meet the editor

Hongliang Luo obtained his bachelor's and master's degrees at the School of Mechanical Engineering, Yanshan University, China in 2012 and 2014, respectively. He received his Ph.D. from Hiroshima University, Japan in 2018, and worked as an assistant professor there until 2022. Currently, he is an associate professor at Harbin Engineering University, China. His research interests include biofuel spray impingement, fuel droplet behavior, hydrogen and ammonia combustion, energy, and thermal management. He has published more than forty scientific papers. His current research projects include one national project from China and two from Japan, focusing on the low-carbon energy system applied in powertrains.

Contents

Preface

Droplet dynamics can be used in many industrial and technical applications, such as coating machines, cooling systems, gas turbines, spray painting, nozzle baffle valves in hydraulic systems, and combustion engines in marine and aircraft vehicles. Different applications require different solutions and effects. In small combustion engines, such as direct injection spark-ignition (DISI) gasoline commercial vehicles, impingement of atomized droplets on the piston head and cylinder wall affects the mixture formation prior to combustion and thus influences engine performance and pollutant emissions. Especially during the warmup or cold start condition, the droplets impact the wall to form liquid adhesion before evaporation, resulting in unburned hydrocarbons. Even in marine and aircraft engines with less consideration of emissions, droplets impacting and adhering to the wall may decrease the efficiency and performance of the engine, leading to more fuel storage. As such, droplet behavior is a basic scientific and critical issue for investigation. Furthermore, mastering the principles of droplet dynamics helps to optimize simulation models and improve computational efficiency and accuracy.

This book on fundamental research and application of droplet dynamics includes six chapters that address such topics as droplet behaviors, experimental methods for investigations, multiple droplet characteristics, liquid transportation in atmospheric pressure plasmas, liquid spray and atomization in a nozzle, and droplets in bioinspired smart surfaces. In practice, these aspects must be considered together to maximize the understanding and application of droplet dynamics. The chapters are organized into four sections on concepts, investigation methods, fundamental research, and applications.

A comprehensive analysis of the technology and application of droplet dynamics is beyond the scope of this book. However, the content of this volume is useful for audiences to broaden their knowledge of liquid droplet behaviors. This book may serve both as a graduate-level textbook for mechanical engineering students and as a reference for professional engineers in the industry.

The research necessary to write and edit this book was carried out during my employment as an assistant professor at Hiroshima University, Japan. I am grateful for my experiences there as well as the support of my colleagues. I would also like to thank IntechOpen for inviting me to be the editor of this volume, especially Ms. Ana Javor and the publishing process staff for their help in coordinating the reviews, editing, and printing of this book.

Dr. Hongliang Luo
College of Power and Energy Engineering,
Harbin Engineering University,
Harbin, China

Section 1

Introduction

Chapter 1

Introductory Chapter: Droplet Formation and Evolution

Hongliang Luo

1. Introduction

Nowadays, decreasing carbon emissions becomes the global consensus. Therefore, in order to achieve carbon neutrality in the near future, many efforts should be done including energy transition, carbon capture, and carbon utilization. Among them, liquid-droplet flow can be applied in many industries, such as the internal combustion engine, colling machines, coating machines, hydraulic transmission equipment agricultural irrigation, and oil-gas transportation.

2. Liquid droplet formation

Generally, nozzles or orifices are often applied to disperse the liquid into the air environment or another immiscible liquid. The discrete droplet is called the discrete phase, while the gas or other liquid is called the continuous phase. In addition, during the liquid-gas interaction, the liquid film may still break into small droplets. Therefore, in the industrial field, especially in the field of internal combustion engines, the discrete phase and continuous phase fluids move together, finally forming a common two-phase (gas-liquid) fluid. For example, the liquid fuel is firstly injected into the cylinder and atomized by the air movement. After fully mixed with air, fuel droplets are ignited and then explosively burned. The main mechanical behaviors of droplets are shown as follows:

1. "Internal circulating flow"—The shear force generated by the friction between the two phases in the continuous phase fluid causes the droplets to flow, called internal circulating flow.

2. "Deformation"—Small droplets are spherical, while large droplets tend to deform and eventually become ellipsoids due to uneven pressure distribution on the droplet surface.

3. "Oscillation"—When the Reynolds number of the droplet is large, the behavior of surface vibration and even vibration deformation will occur, called oscillation.

4. "Breakup"—The droplet itself breaks into several droplets, or multiple droplets collide and then break into several droplets.

5. "Coalescence"—When the droplets collide with each other, they merge into larger droplets due to viscous forces.

IntechOpen

3. Impingement and evaporation

After droplet formation, it moves forward and may impact the solid wall or other phase, some behaviors then can be involved as shown in **Figure 1**.

1. "Stick"—in which the impinging droplet adheres to the wall in nearly spherical form. This occurs when the impact energy is very low and the wall temperature T_w is below T_{PA} (pure adhesion temperature, below it adhesion occurs at low impact energy).

2. "Spread"—where the droplet impacts with a moderate velocity onto a dry or wetted wall and spreads out to form a wall film for a drywall, or merges with the pre-existing liquid film for a wetted wall.

3. "Rebound"—in which the impinging droplet bounces off the wall after impact. This regime is observed for two cases: (a) on a drywall when $T_w \geq T_{PR}$, (pure rebound temperature, above which bounce occurs at low impact energy), in this case, contact between the liquid droplet and the hot surface is prevented by the intervening vapor film; (b) on a wetted wall, when the impact energy is low, and the air film trapped between the droplet and the liquid film causes low energy loss and results in bouncing.

4. "Rebound with breakup"—where the droplet bounces off a hot surface ($T_w < T_{PR}$), accompanied by break up into two or three droplets.

5. "Boiling-induced breakup"—in which the droplet, even at very low collision energy, disintegrates due to rapid liquid boiling on a hot wall whose temperature lies near the Nakayama temperature T_N (is the Nakayama temperature at which a droplet reaches its maximum evaporation rate).

6. "Breakup"—where the droplet first undergoes a large deformation to form a radial film on the 'hot' surface ($T_w > T_{PA}$), then the thermo-induced instability within the film causes the fragmentation of the liquid film in a random manner.

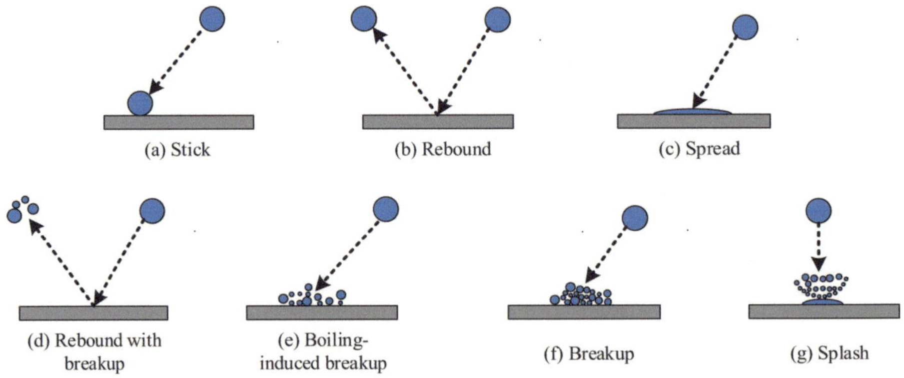

Figure 1.
Droplet impacting behaviors.

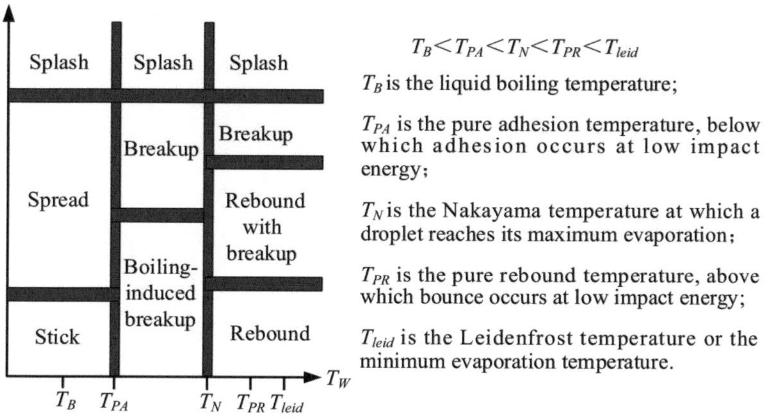

Figure 2.
Behavior transition conditions.

7. "Splash"—in which, following the collision of a droplet with a surface at very high impact energy, a crown is formed, jets develop on the periphery of the crown and the jets become unstable and break up into many fragments.

The existence of these impingement behaviors is governed by a number of parameters characterizing the impingement conditions. These include incident droplet velocity, size, temperature, incidence angle, fluid properties such as viscosity, surface tension, wall temperature, surface roughness, and if present wall film thickness and gas boundary layer characteristics in the near-wall region. Quantitative criteria for the behavior transitions from Bai and Gosman [1] and refined in the present work are presented in **Figure 2**.

All these droplet behavior including formation, evaporation, and evolution should be clarified to clearly understand the droplet dynamic. Especially for the current "carbon cycle" age, all the equipment should be re-design or developed with less CO_2 emission to protect local environments. Among them, the droplets dynamic can be applied in many new technologies or even develop future renewable fuels.

Acknowledgements

The author would like to acknowledge the National Natural Science Foundation of China [Grant 51909037] and the Foundation of State Key Laboratory of Engines [No. K2022-12].

Conflict of interest

The authors declare no conflict of interest.

Author details

Hongliang Luo[1,2]

1 Foshan University, Foshan, China

2 Hiroshima University, Higashi-Hiroshima, Japan

*Address all correspondence to: luo@hiroshima-u.ac.jp

References

[1] Bai C, Gosman AD. Development of methodology for spray impingement simulation. Journal of Engines. 1995;**104**(3): 550-568

Section 2

Investigation Method

Chapter 2

Imaging Diagnostics for Jet Breakup into Droplets: A Review

Anu Osta

Abstract

A concise review of the recent developments in some of the standard optical diagnostics applied for primary jet breakup studies has been presented here. Primary breakup is the core breakup of liquid jets and sheets into droplets upon its interaction with the ambient gaseous atmosphere. This phenomenon is encountered in various aerodynamic, fluid dynamic, and combustion situations. The imaging diagnostics reviewed here include photography, high-speed imaging, shadowgraphy, digital holography, ballistic imaging, jet core illumination, thermal imaging, Mie imaging, x-ray phase contrast imaging, and laser-induced fluorescence. The advantages and limitations of each technique, their success, and future developmental trend are discussed.

Keywords: atomization, diagnostics, liquid-jet, visualization

1. Introduction

Liquid atomization is a phenomenon associated with liquid fuel combustion processes, industrial and agricultural sprays, and our daily life activities such as when using body or hair sprays. When a liquid column or a sheet issuing from a source (nozzle or channel), interacts with the ambient atmosphere instabilities develop inside the liquid sheet or the column core. Instabilities also develop at the atmosphere-liquid interface in the form of flow structures like surface waves and ligaments. Flow conditions prevailing inside the source boundary such as void cavities or turbulence also affect instability development and the resulting bulk liquid disintegration. **Figure 1** depicts typical atomization in a liquid jet that is subjected to a crossflowing fluid like air. The liquid disintegration is also referred as the 'primary breakup' process. Primary jet breakup is often regarded as the first step in a jet atomization process. The instabilities are mostly of the Kelvin–Helmholtz (KH) and Rayleigh Taylor (RT) type. The primary droplets formed in the initial stage of atomization may then undergo subsequent secondary breakup stages.

Complete characterization of spray atomization would require one to analyze it at both the macroscopic and microscopic scales. Macroscopic characteristics would reveal the spray volume and its penetration into the surrounding atmosphere (both axially and radially), spray cone angle, mass flow rate, spray momentum flux, and the mass distribution of the spray fluid further downstream after atomization. The microscopic properties would reveal droplet sizes, droplet velocities, droplet number density, droplet distribution, and the temperature field of the spray.

IntechOpen

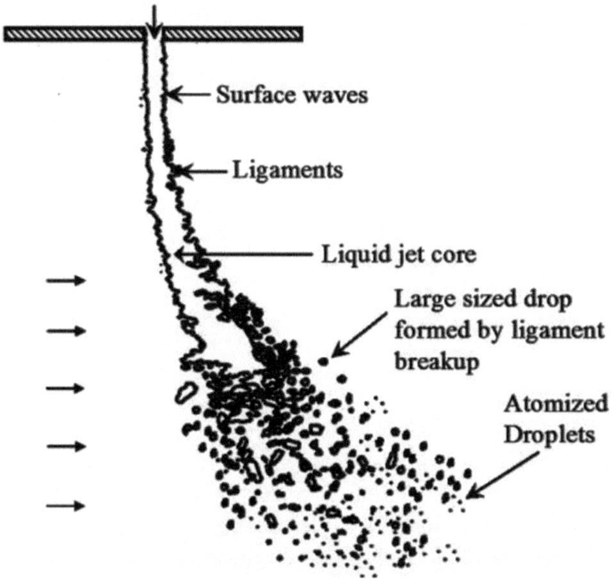

Figure 1.
Schematic representation of primary breakup of a liquid jet in crossflow.

Over the past few decades, various techniques [1–5] have developed both intrusive and non-intrusive to visualize the primary breakup process. These are based on physical jet interaction, thermal response, electrical response, and optical imaging techniques. No single diagnostic can completely characterize the entire spray structure. A combination of diagnostics is often applied to obtain detailed information about the spray characteristics. Newer developments in the field of optics, and electronics have vastly improved the traditional imaging and probing techniques yielding significantly better results.

The effectiveness of any visualization technique depends on several factors. The optical setup, illumination quality, light source, dynamic range and spatial resolution, light sensitivity, frame speed, and signal-to-noise ratio of the camera sensors – all play an important role. Care should be exercised to minimize the errors associated with non-uniform or unstable illumination, curvature effects, reflections, and shadows. When dealing simultaneously with structurally different entities like liquid core, liquid surface structures, and spatial droplet distribution, various challenges present themselves. They include the construction of three-dimensional atomization map from two-dimensional images, optical inaccessibility of dense jet breakup region, high-speed imaging without sacrificing high resolution, signal loss due to high noise, diffraction blurring of small droplets, overcoming multiple scattering, optical aberrations, and attenuation to name a few. The rest of this chapter will discuss some standard techniques for visualizing the primary breakup region.

2. Light sources and imaging devices

All optical techniques are based on certain lighting schemes, such as direct lighting, diffused lighting, flood lighting, trans-illumination, reflective, etc. In the *trans-illumination scheme,* a light beam is passed through the spray or sheet and is imaged on

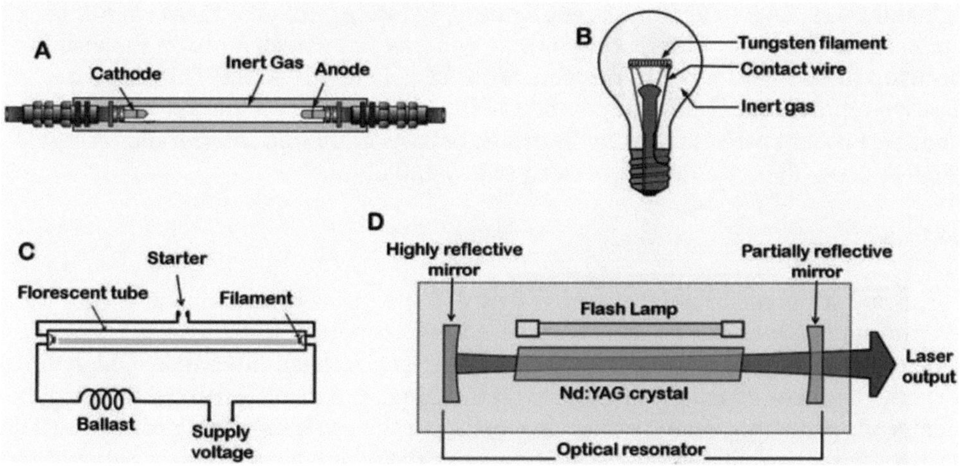

Figure 2.
Some of the commonly used light sources for imaging purposes. A-gas arc lamp used as stroboscopic source, B - incandescent light source, C- fluorescent light source, D - Nd:YAG laser light source.

the other side. In the *internally illuminated scheme,* a light source is located inside the spray and light is transmitted through the spray core to the surrounding. In a *reflective scheme,* the incident light reflects off the liquid surface, while in *fluorescence*, a thin light sheet illuminates a planar section of the spray. The light sources used for illumination could be either coherent or non-coherent type (**Figure 2**). Incoherent sources include strobe, incandescent, halogen, arc, and fluorescent lamps among others. Coherent sources include lasers and LEDs.

2.1 Strobe light

Strobe lighting involves producing flashes of light for a short period of time at regular intervals. The typical flash lasts for 200 μs and may be synchronized with the framing rate of a suitable camera. With stroboscopic illumination, one can prolong the source lifetime, operate at increased light intensities, 'freeze' the motion of a fast-moving object, and time the pulses.

2.2 Incandescent light

Incandescent lighting works by producing light by heating a wire filament. They have a wavelength of 300 nm – 1500 nm, flicker between 60 and 120 Hz, have an orange-yellow color cast, and are considered harsh for imaging purposes. They used to be widely used in the past for scientific imaging owing to their color being similar to natural sunlight. However low lamp efficacy (lumens per watt), luminaire efficiency, poor controllability of the light source, and incoherency presented challenges. Tungsten filament lamps, Halogen lamps, and Xenon arc lamps are some examples of incandescent light sources.

2.3 Fluorescent light source

When electricity is passed through mercury vapor in a glass tube the radiation emitted interacts with the glass coating to produce white light. The illumination is high and this is known as fluorescent lighting. However, most of them cannot be dimmed and they may Fluorescent light commonly has wavelengths in the visible spectrum of 400–700 nm and frequencies of 10 kHz to 100 MHz. They have a negative reputation in photography due to the blue or green color cast that they produce. Flicker correction using electronic ballasts and radio interference removal using suitable filter circuits might need to be applied.

2.4 Lasers

Lasers have come to be regarded widely as the most versatile source of light for imaging applications due to their many desirable properties like high brightness, stability, longevity, narrow spectral bandwidth, narrow beam divergence, high degree of spatial and temporal coherence, and well-defined polarization properties. A disadvantage of using laser beams is that they exhibit a Gaussian intensity profile, revealing particles in the middle region while particles at the edges get concealed. Some of the common types of lasers used in laboratories for research purposes have been discussed below.

2.4.1 Solid state lasers

These are one of the most commonly used laser systems, e.g. Nd:YAG Laser. They operate in the infrared 1064 nm wavelength regime, but pulses can be frequency doubled to generate wavelengths at 532 nm, or higher harmonics, 355 nm/266 nm. They can be operated in both the pulsed and the continuous mode and have an average power density of 5×10^{-3} W/cm^2/10s.

2.4.2 Gas lasers

In gas lasers, an electric current is discharged through a gas or plasma to produce coherent light in the ultraviolet (excimer or nitrogen lasers) and visible range (He-Ne or ion-gas lasers). The excimer lasers typically operate at 193/248/308/351 nm with a pulse repetition rate of $\sim$100–200 Hz and a pulse duration of $\sim$10 ns. The He-Ne lasers typically operate at 632.8 nm. These lasers can produce beams, with a near-Gaussian/super-Gaussian profile. Molecular gas lasers emit in the infrared region popularly known as infrared (IR) lasers. These lasers can emit between 2 and 1000 μm range, at a rate of 300 GHz and10 THz. Since the optical depth of the jet breakup region in the infrared regime is smaller than in the visible spectrum, the infrared lasers can probe dense regions of the spray more effectively than their visible and ultraviolet counterparts [6].

2.4.3 Ionized-metal vapor lasers

These are an important tool for high-speed flow visualization, e.g. copper vapor lasers. Copper vapors are used as the lasing medium. The emitted pulses are in the green/yellow spectral range (510 nm/578 nm) with a pulse width in the range of 5–60 ns.

2.4.4 Mode-locked lasers

Mode locking or phase locking is a technique used for achieving ultrashort pulses, on the order of picoseconds or femtoseconds. Operating modes periodically interfere constructively to produce an intense burst of light with peak power several orders of magnitude higher than the average power.

Lasers are also classified as continuous wave and pulsed lasers. In continuous wave lasers, the emitted light intensity is constant as a function of time. In pulsed lasers, the energy is released in the form of a short pulse of light.

2.5 LED

LED illumination has recently gained popularity in a number of imaging applications. LED arrays produce intense, even illumination over a given object or area, have relatively low power requirements, generate very little heat, have a very long life and their pulse widths can be finely controlled. LEDs can produce coherent light under specific conditions such as in laser diodes and can be operated in continuous or pulsed mode. Their emissions spectra are broader than that of the lasers.

An integral component of any imaging system is the camera. Almost all of the imaging nowadays is carried out by digital cameras of either the CCD type (charged coupled device) or CMOS type (complementary metal–oxide–semiconductor). A comparison between the two sensor types is presented in **Table 1**. **Figure 3**(a) shows a simplified CCD sensor architecture and **Figure 3**(b) shows a simplified CMOS sensor architecture.

2.6 CCD

A CCD is an image sensor whose working principle is based on the photoelectric effect. It produces electrical charges proportional to the light intensities incident at different locations on the sensor which are then converted to digital values by "shifting" the signals one at a time between stages within the device. A photoactive capacitor array captures a two-dimensional picture of the scene. Each capacitor transfers its charges to its neighbor after which the charge is converted into voltage. The total number of image frames acquired is limited by the on-chip storage capacity. The transfer voltage signals get dampened at very high frequencies. Operating the CCD at very high frame rates leads to heating of the sensor, with accompanying thermal noise. The current practical

Parameter	**CCD**	**CMOS**
Signal-to-noise ratio (SNR)	Higher	Lower
Resolution	Higher	Lower
Repetition rates	Low	High
Speed	Moderate to High	Higher
Cost	Higher cost	Lower cost
Sensitivity	More light sensitive	Less light sensitive
Power usage	Higher	Lower
Dynamic range	High	Moderate
Uniformity	High	Low to medium
Windowing	Limited	Extensive

Table 1.
Comparison of operating characteristics, of CCD and CMOS [7–9].

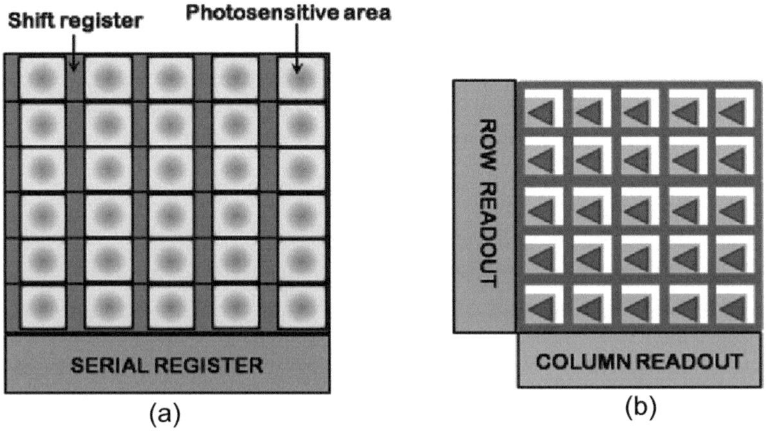

Figure 3.
Schematic of a typical (a) CCD sensor (left) and (b) CMOS sensor architecture (right) [7].

frame rate limit is of the order of 1000 fps which could be extended to up to 100 Mfps (mega frames per second) having a spatial resolution of 312 × 260 pixels per frame [10, 11]. *Intensified CCD* (ICCD) is CCD coupled with an image intensifier for achieving high sensitivity in ultra-low-light-level conditions. They provide better temporal resolution and are suitable for capturing transient events.

2.7 CMOS

A CMOS image sensor consists of an integrated circuit containing an array of pixel sensors, each pixel containing a photodetector and an active amplifier. Each pixel can be read individually. This enables fast clocking speeds (time taken to read the charge of the sensor) and high frame rates. The CMOS devices are highly immune to noise and have low static power consumption. Since the CMOS sensors have readout transistors at every pixel, most of the photons falling on the chip hit the transistors instead of the photodiode, lowering the light sensitivity of the CMOS chip. CMOS is more suited for faint/low light conditions and requires taking longer exposures. Some of the latest CMOS cameras can reach frame rates of 285,000 fps at reduced resolution[1] or the high definition[2], 1080 HD resolution up to 2000 fps.

In order to effectively visualize the liquid breakup region under various illumination and density constraints, it is important to have a good understanding of the essential basics of *photography*. Some of these parameters e.g. field of view, frame rate, exposure, aperture, magnification, depth of field, depth of focus, and the dynamic range shown in **Figure 4**, are discussed below.

2.8 Field of view

The field of view in the case of optical instruments is the solid angle through which the camera is sensitive to light. It defines the area that the camera is able to record, and is a function of the working distance, the focal length of the lens, and the sensor/film area.

---

[1] MotionBLITZ EoSens® mini2

[2] Photron Fastcam BC2

DOI: http://dx.doi.org/10.5772/intechopen.107370

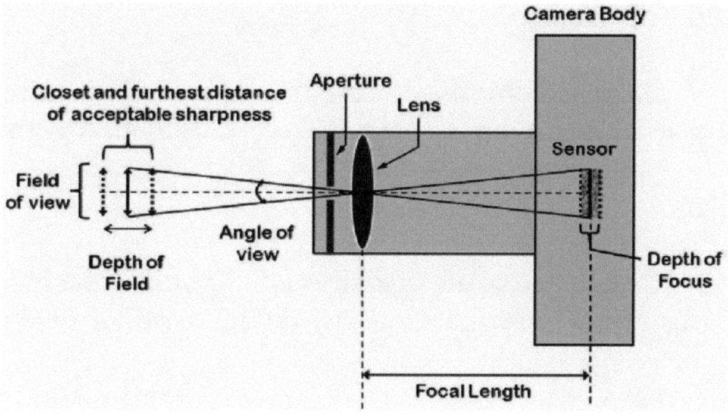

Figure 4.
Schematic of a typical camera and the different imaging parameters.

2.9 Frame rate

Frame rate is the number of consecutive images recorded per unit time by an imaging device. It is mostly expressed in frames per second (FPS) or hertz (Hz). For a fast event like atomization, a low frame rate may result in jerky and less fluidic motion while a very high frame rate would require high processing power and storage to yield sharp high-quality images of the phenomena in motion.

2.10 Exposure

It is the amount of light collected by the sensor of the camera during single image capture. Too long exposures lead to excessive light collecting on the film, which results in a washout appearance. Too short exposures lead to insufficient light collection resulting in a dark image. Exposure depends on the frame rate and camera aperture. The exposure time in CCD and CMOS cameras is set by an electronic shutter that is controlled either manually, electronically, or by software.

2.11 Aperture

Aperture is the size of the lens opening which limits the amount of light entering the camera and falling on the image plane. It controls the depth-of-field in the sense that the background can be blurred with a wide aperture keeping just the object in focus or alternatively keeping everything in focus by using a narrow aperture. A wide aperture results in a higher degree of optical aberrations (distortions) and vignetting (falling intensity toward the edges of the picture).

2.12 Magnification

Magnification is the degree of scaling (enlarging or diminishing) of a subject on the image plane. There are two ways to represent magnification, (a) Linear or transverse magnification - it is represented by Y/X where Y is the image length and X the subject length and (b) Angular magnification - angle subtended by the object at the focal point.

2.13 Depth of field

It is the range of distance along the optical axis between the nearest and farthest objects in a scene that appear to be in focus (sharp) in the photograph.

2.14 Depth of focus

This is the limit of the image plane displacement at which the image will appear sharp. Depth of focus refers to image space, while the depth of field refers to object space.

2.15 Dynamic range

It is the ratio of the saturation level of a pixel to its signal threshold or simply put the ratio between the maximum and minimum measurable light intensities. Since the dynamic range of sensors is far less than the human eye, local tone mapping and dynamic range adjustment are used.

The *spatial resolution* of an imaging system is its ability to distinguish separate objects within its field of view. For the same sensor size, increasing inthe field of view decreases the image resolution.

Illumination setup plays a vital role in determining the clarity of an image. Various lighting techniques exist [5]. The most commonly used method is frontal lighting in which the camera and illuminating source are placed in front of the jet and the light reflected from the breakup entities is recorded. *Frontal lightening* gives a three-dimensional appearance that helps in visualizing the liquid surface features. This arrangement fails to yield any inner details of the jet breakup. Illumination of a section of the liquid by a light sheet is followed mostly in cases of axisymmetric flows. Weak elastic scattering by the droplets at the breakup location helps to reveal the inner details of the breakup location in the plane of illumination. However in this case, multiple scattering is often a major drawback. *Backlighting*, is a technique in which the light source is placed behind the object, and translucent glass is located between the object and the light source. The translucent screen diffuses the flash uniformly over a wider region illuminating the entire breakup location section. This diffuse light can produce diffuse reflections from the object being backlit resulting in soft and blurry edges. *Bidirectional lighting* is when two light sources are placed at an angle of 120° from the camera's line of sight, illuminating the breakup location.

3. Diagnostics

Different optical techniques are used to characterize the primary breakup process; e.g. photography, shadowgraphy, holography, ballistic imaging, jet core illumination, laser-induced fluorescence, thermal imaging, Mie imaging, and X-ray phase contrast imaging. **Table 2** summarizes the application situations of the different diagnostics and the type of information they provide.

3.1 Photography

Photography in its simple form consists of an illumination source illuminating the object and a camera recording the images of the object. Photography of a jet breakup is

Diagnostic	Underlying feature	Type of breakup	Information	Limitations
Photography	Reflection-jet is illuminated by an external light source and imaged.	Any	Image of the focused jet surface (ligaments, drops, liquid core).	The resolution, depth of field, frame speed, and 2D.
Shadowgraphy	Transmission- projected shadow intensity of the jet periphery is imaged.	Less to moderately dense jet breakups	Jet peripheral structures (ligaments, drops, liquid core).	No information about the frontal and back-ward jet surface structures. Information loss due to shadow overlap. 2D.
Holography	Interference between reflected/refracted wave fields from the object and a coherent reference wave is imaged and then reconstructed.	Less to moderately dense jet breakups	3D information in form of 2D,	Information loss due to diffraction pattern overlap. Low depth of focus
Ballistic imaging	Transmission- ballistic photons are imaged.	Dense jet breakups	Spray structures inside the optically dense atomized regime as well as on the surface.	Opaque to the jet core structures. Limited resolution due to large scattering. 2D.
Jet core illumination	Light propagates through the liquid jet core which fluoresces when a fluorescing dye is added to the liquid.	Any	Breakup location, elements of breakup region, liquid jet surface features.	Not easily adaptable, scattering high for dense breakup regions, no information on droplet size or velocity, 2D.
Laser induced fluorescence	Liquid jet illuminated by a laser, fluoresces.	Less to moderately dense jet breakups	Jet breakup structures in a specific planar section of the jet and the surrounding vapor.	Scattering, non-uniform intensity distribution, attenuation, interference from other species, 2D, weaker than Mie signal.
Mie imaging	Light scattering signal from jet breakup particles having the size of the order of the scattered wavelength of light.	Dense jet breakups	Jet breakup structures in a specific planar section of the jet, particle size.	Scattered and refracted rays interfere leading to a ripple effect, assumes a spherical particle.
X-ray phase contrast imaging	Contrast due to the phase shift undergone by the beam upon its passage through the sample because of photon absorption and photon scattering.	Dense jet breakups	Three-dimensional jet surface topology projected on a two-dimensional image, structures inside the jet core.	Inability to distinguish the front and back surfaces.
Thermal imaging	Temperature visualization - laser-induced emission of thermographic phosphors or organic tracers (fluorescence and phosphorescence).	Any	Jet breakup structures in a specific planar section of the jet and the surrounding vapor.	Quenching, cross-talk, optical thickness, temperature gradient, attenuation, scattering, 2D.

Table 2.
A comparison of the different diagnostics available for visualizing primary breakup.

used for getting measurements on breakup lengths, drop sizes, droplet distribution density, jet and droplet velocities, fluid flow behavior, etc. In some cases, a diffuse screen may help scatter the light incident on the object, and digital display help in visualizing the image in real-time as shown in **Figure 5**. Imaging could be in the single-shot, low speed, and high-speed modes. Illumination could be provided by any of the schemes discussed previously, e.g. strobe synchronized with the camera frame rate in the forward light scattering configuration [12–14]. Incandescent or pulsed laser sources could also be used.

Figure 6 shows a high-speed photograph of a jet surface undergoing primary breakup in still air [15] where the jet was back-lighted at 45° from the horizontal and shielded from ambient light (**Figure 7**).

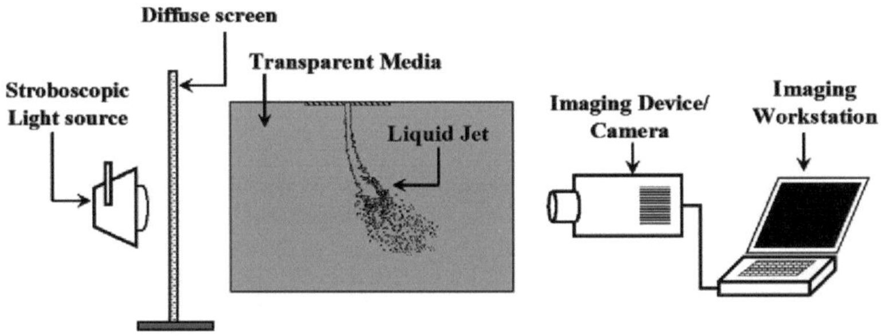

Figure 5.
Optical setup of the photography technique.

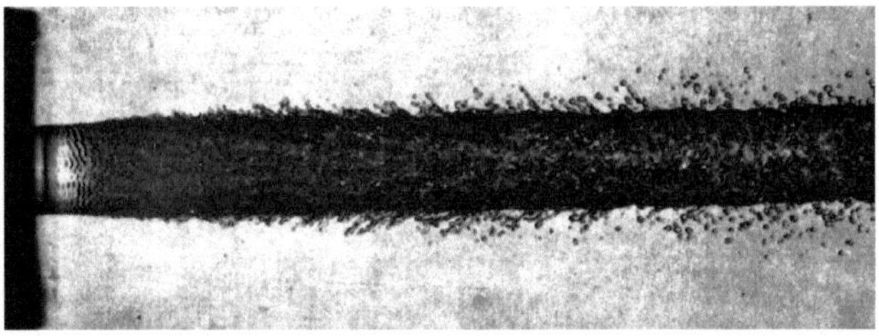

Figure 6.
Photographic image of jet surface primary breakup in still air [15].

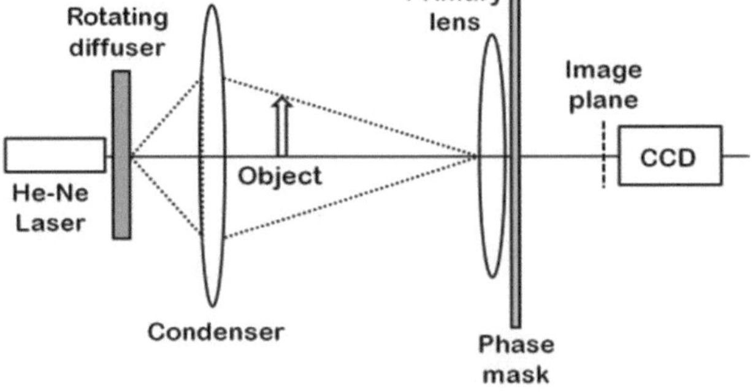

Figure 7.
Laser backlit illumination [16].

With increasingly dense sprays, the intensity of the back-lit illumination should increase to distinctly reveal the droplets in the breakup location. Magnification is pivotal in distinguishing and measurement of atomization entities. Droplets, ligaments, and surface features in the size range of micrometers would require significant magnification. Higher magnification comes at the expense of the depth of field and this presents a challenge toward achieving a focused image of the small non-stationary liquid surface. Optical systems have been designed to make it flexible such as by using a phase mask [16] consisting of a combination of Fresnel lenses (FL) (**Figure 8**) where each FL works in tandem with the primary lens to produce a sharp image for a unique object plane. A micro-actuator can be used to translate the detector along the optical axis during image integration [18]. Image processing algorithms [19] applying the focus stacking method can modify the phase of incoherent light wavefront to produce a point-spread-function (PSF) over a large region of focus yielding an extended depth-of-field. For visualization of the breakup phenomena over a wide range of distances, a large field of view with a high depth of focus might be desirable when the magnification is low.

Spatial resolution is often denoted by lines per inch (lpi) or μm. It represents how closely two lines can be resolved in an image. Film-based photography has a spatial resolution of ~100 μm, while digital photographs could achieve resolutions up to 10 μm. Digital image resolution is limited by pixel noise and pixel cross-talk. The spatial resolution also depends on the conical angle subtended by the object at the lens aperture. Lens resolution for narrow lens aperture is affected by diffraction, and for large apertures affected by optical aberration. For imaging small structures at widely separated points with a high resolution, a multiple-segment long-distance microscope in combination with micro-lens and aperture array has been demonstrated [17] (**Figure 8**).

The finite size of a camera lens with a circular aperture leads to the diffraction of parallel light rays passing through them to form a diffraction pattern in the image. It usually has an appearance of a central bright spot and surrounding bright rings, separated by dark nulls. This two-dimensional far-field diffraction pattern is called '*Airy disc*', (**Figure 9.**). Its angular radius measured from the center to the first null is equal to $\sin^{-1}(1.22\lambda/D)$, where λ is the wavelength of light passing through and D is the aperture diameter. The diameter of the first dark circle (width of the disk) defines the theoretical maximum resolution for an optical system such that if two objects imaged by the lens start to have overlapping airy discs by the virtue of their closeness, a blurring effect occurs.

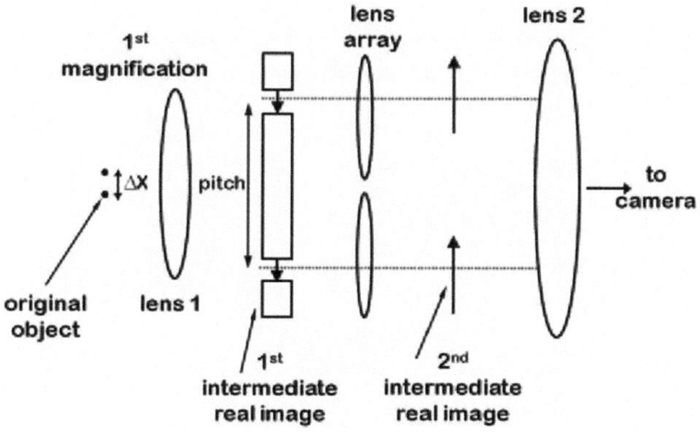

Figure 8.
Optical setup of Bauer [17].

Figure 9.
Airy disc.

Figure 10.
Tsai calibration grid.

Camera calibration is done both geometrically and photometrically. Geometric calibration involves using a Tsai grid (two planes at the right angle with checkerboard patterns as in **Figure 10**) by which the scaling factors between the image and the actual target dimensions are achieved. In photometric calibration relation between digital counts and luminance is sought by capturing test patches with known relative luminance. Photography also has some drawbacks. Mainly its inability to resolve a three-dimensional perspective of the object accurately, inability to see through optically dense droplets or liquid structures, and limitations with respect to depth of focus/magnification. For cameras, high spatial resolution comes at the expense of reduced pixel size and therefore a reduced light-sensitive area and reduced signal-to-noise ratios. The absence of cameras and light sources with very high repetition rates, high enough to capture fast motion continuity is another limitation in photography. **Figure 11** shows a setup [20]

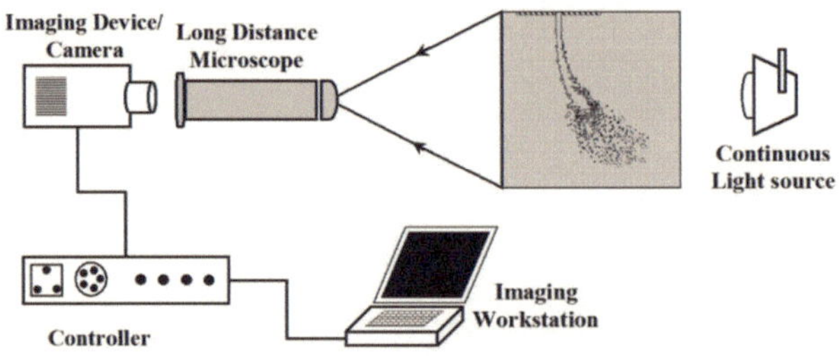

Figure 11.
High magnification video setup [20].

for obtaining high-magnification video images of a jet breakup. In this setup, a high-speed CCD camera fitted with a long-distance microscope lens faces the spray which is itself illuminated with a flat-faced halogen lamp from the rear. A computer records and stores the high-speed movie at the rate of 2000 frames/sec. The field of view, in this case, was 13.9 mm x 8.9 mm, with an image resolution of 29.6 pixels/mm.

3.2 Shadow imaging

Shadowgraphy is a technique of imaging the shadow of the refractive index field. When light passes through a region of varying refractive index (**Figure 12**), it experiences retardation proportional to the material's density and bends toward the region of higher refractive index. The angular deflection and displacement of the rays are small. The local beam intensity is not significantly affected, but the angular deviation is enough to produce a focusing effect above the higher refractive index regions as the beam propagates beyond the fluid layer. This coupled with the retardation causes the wavefronts to turn, due to which rays converge and diverge into bright and dark regions.

The refractive deflection of rays causes a shadow effect (spatial modulation of the light-intensity distribution) in the recording plane which is then imaged. The image intensity thus depends on the variations in the optical density of the object media. A portion of incident light refracts at the fluid-interface boundary and may produce a darker boundary region. The portion that does not interact with the object produces a homogenous background. The dark regions (shadow) in a shadowgram mark the boundaries of the object. Light transmittance is affected by scattering and absorption too.

Figure 13 shows a shadowgraph image [21] of quenching oil jet spray from a 1.2 mm atomizer nozzle at 15 m/s and 38°C. Shadowgraphy has been used to visualize liquid jet breakup [23, 24] at different length scales in-order to determine ligament-droplet sizes, and speeds. Shadowgraphs can be subjected to further image analysis for detecting the liquid contour and its individual features based on the comparison of the RGB (Red Green Blue) intensity levels to a pre-established threshold level or identification of local changes of RGB intensity, which are their greatest at the fluid boundaries. Shadowgraphy is different from backlit photography despite a light source being located behind the object in both cases. For backlit photography, the background lighting is diffusivein nature. In addition, the liquid frontal surface is illuminated and the reflected light is directly photographed with a camera. The shadow image formed

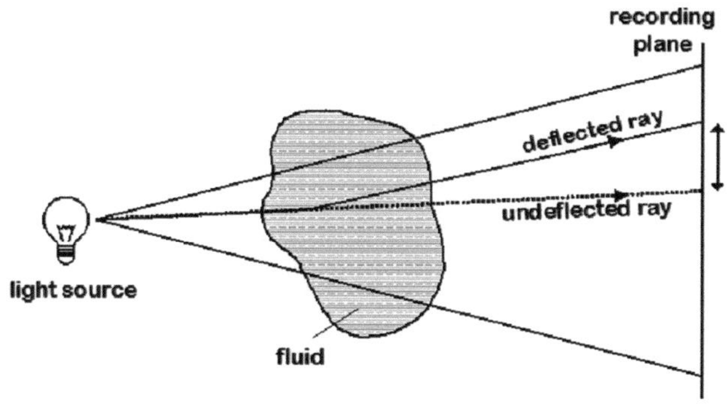

Figure 12.
Formation of shadow image by relative deflection of rays.

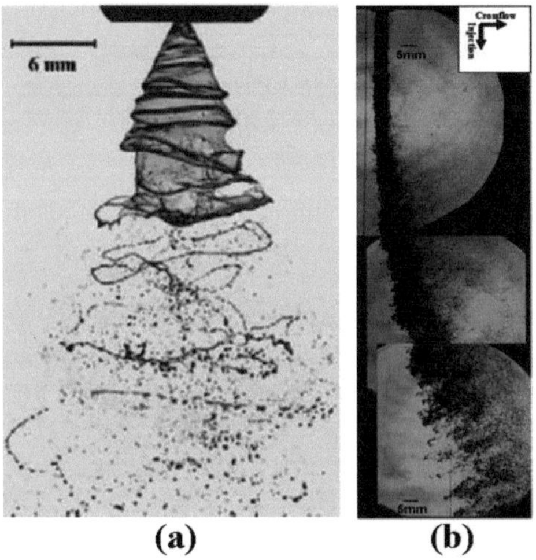

Figure 13.
Shadowgrams of (a) quenching oil jet spray [21] and (b) water jet spray [22].

on a screen is imaged by the camera. For optically dense objects the intensity variation inside the image is insignificant compared to the surrounding bright background intensity. In such cases, a dark shadow image with a bright background may result and therefore any kind of density variations analysis would be irrelevant. Shadowgraphs provide phase information (refractive index gradients) [21, 25] or simply a monochromatic projected shadow image of the object [22, 26–28]. It can help in visualizing convective flows [29].

Figure 14 shows a typical laboratory setup using incident parallel laser light forming a focused shadowgraph image onto a screen. It is known as "focused shadowgraphy". Light from the source is passed through a spatial filter, then collimated by using a collimating lens, and allowed to pass through the object. After passage, the beam is focused using a relay lens which forms a real inverted image onto a translucent or a ground glass screen or a photographic film. The camera is focused on this image from the other side of the translucent screen. Alternatively, the primary shadowgraph formed by the relay lens can be recorded directly with a camera simply by focusing the camera lens on the plane of the primary shadowgram. This allows variable magnification to be achieved as well. The relay lens somewhat limits the field of view of focused shadowgraph but sizes to the order of micro scales can be obtained using the appropriate magnification. The use of photographic films yields better resolution [26, 30, 31].

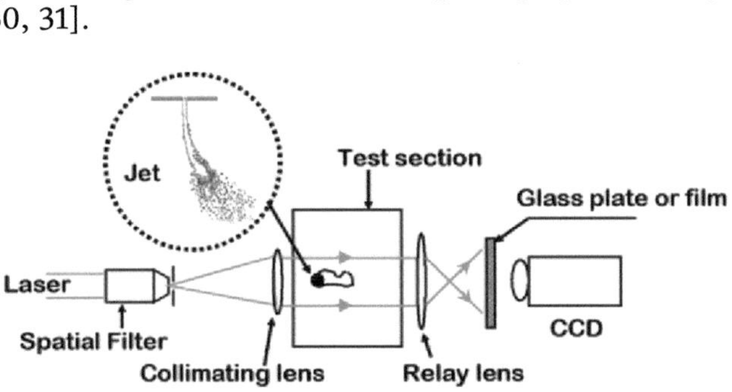

Figure 14.
'Focused shadowgraphy' setup (shadowgraph is formed on glass plate or film).

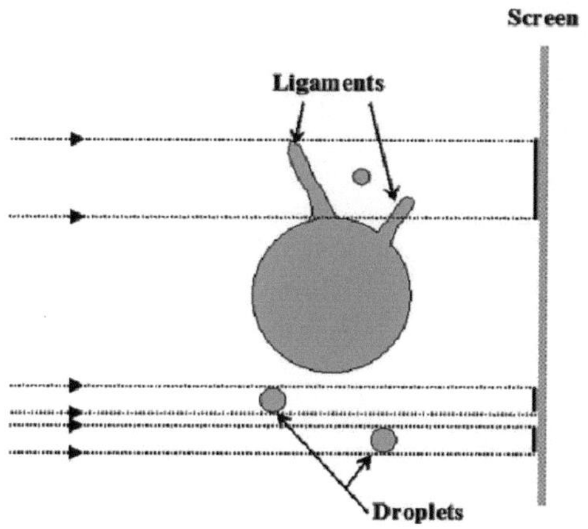

Figure 15.
Shadowgraph's inability to image entities hidden or obscured by the projection of the larger structures.

Shadowgraphy is easy and inexpensive to carry out, has a high spatial resolution for most practical purposes, is independent of the shape and material of the phase media and its relatively large field of view is well suited for size, shape, position, and velocity determination.

A disadvantage of shadowgraphy is that since it gives the projected image of the object, any overlapping ligament or droplet information is lost if its projection does not fall along the jet periphery or is obscured by a larger structure in its line of sight (see **Figure 15**). It fails to give details of the inner fluid structure of the primary breakup region high-density region. It is also unable to distinguish between fuel vapor and small droplets. Image sensitivity, blur, and feature size are always competing issues in shadowgraphy which makes precise measurements difficult. The formation of caustics in shadowgrams is a particular disadvantage when the caustics become confused with some other phenomenon. For instance, a boundary layer can act as a cylindrical lens, focusing light into a bright line.

Basic shadowgraphy has been adapted to other imaging techniques such as Microscopic shadowgraphy, Stereoscopic shadowgraphy, and Holographic shadowgraphy [32].

Spark-shadowgraph technique [33, 34], High-speed shadowgraphy (HSS) [35, 36], and Specialized Imaging Shadowgraph (SIS) are some of the other related developments.

The tomographic Shadowgraphy technique [37] is based on a multiple-view camera setup and is capable of resolving the liquid jet core both spatially and temporally (**Figure 16**). The spray shadowgraphs are obtained using a pulsed LED and four synchronized double frame cameras angled 30° to each other. The spray is then reconstructed by using a line-of-sight reconstruction technique. Stereoscopic shadowgraphy [38] uses four parabolic mirrors to form two inclined intersecting beams and the object is placed at the intersection. Two cameras synchronized in the master-slave configuration record the shadowgraph pairs automatically. Upon image reconstruction, three-dimensionality is achieved.

3.3 Holography

Holograms are made by the interference between a wave field scattered from the object and a coherent background, called the reference wave [39–41]. They can focus

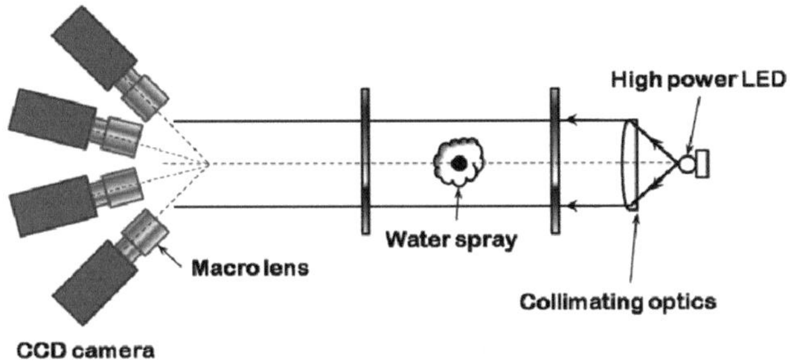

Figure 16.
Tomography shadowgraphy setup [37].

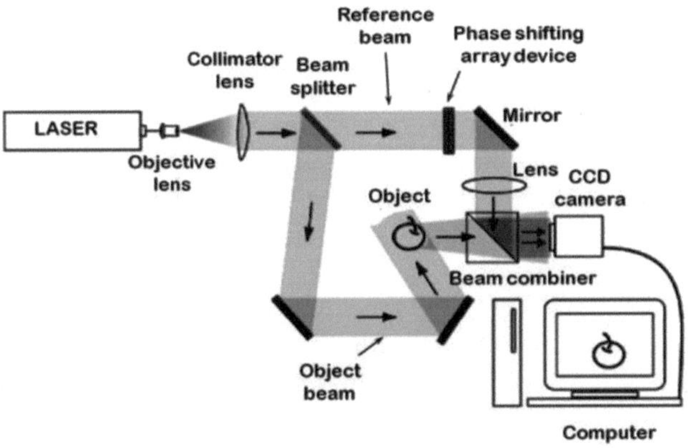

Figure 17.
The optical system of parallel phase-shifting digital holography [43].

simultaneously in three-dimensional spaces [42]. A sketch of a digital in-line holography setup is shown in **Figure 17**. The optical setup consists of an incident light source (laser). The incident beam is collimated using a beam expander (objective lens + spatial filter) and convex lens combination followed by splitting it into the object and the reference beams using a beam splitter. The object beam is passed through the object where it suffers diffraction and a phase change. Both the reference and the object beams are adjusted for intensity before being combined via a beam splitter. The resulting interference pattern is then recorded on a CCD sensor at the hologram plane which is at a finite distance from the true image plane. Magnification can be introduced by using a convex lens as a relay lens after the object beam has passed through, the object, and this magnified hologram would then be recorded on the CCD sensor. The recorded hologram is then reconstructed using a computer program based on the convolution type approach which solves the Rayleigh Sommerfeld formula for the reconstruction of a wave field [40, 44] using the Fast Fourier Transform algorithm (or the Fresnel-Kirchhoff for high numerical aperture or any of the other Fraunhofer transform/wavelet transform schemes).

Holography eases the limitation of the depth of field and provides a three-dimensional image made up of two-dimensional image planes focused at different distances along the axis perpendicular to the image plane [45]. Unaffected by the non-spherical droplets and ligaments that are usually encountered very close to the injector exit, it also has a virtually unlimited depth of focus, and high magnification even for a

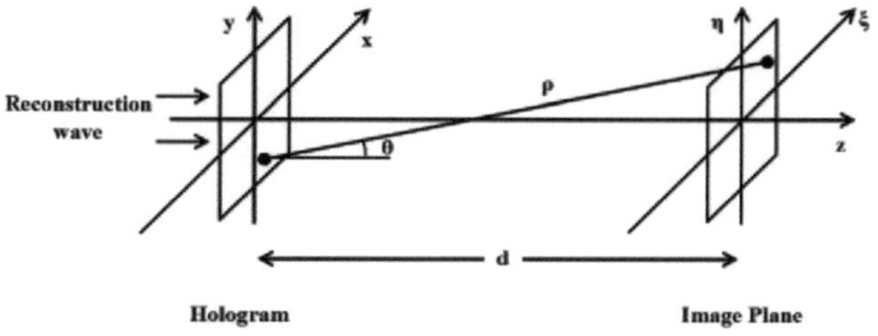

Figure 18.
Hologram reconstruction coordinate system.

small field of view. Depending on the angle between the reference wave and the object wave, a zero reference angle setup is called inline holography and a non-zero reference angle setup is called an off-axis holography. They can be recorded using a photographic film [46], or digitally using a CCD [47–49]. Though film-based holography provides a high spatial resolution the time involved in carrying out the wet processing of film makes it disadvantageous to be employed for measuring dynamic phenomena.

Digital holography enables quick recording and retrieval of hologram information in real time. In **Figure 18**, ρ is the distance between a point in the hologram plane and a point in the reconstruction plane, ξ, η are coordinates in the reconstruction image plane, and x, y are coordinates in the recording hologram plane. The hologram is reconstructed to different spanwise distances so as to focus on the particles centered at those distances from the camera in the jet breakup field. Different modifications to the reconstruction process have been adopted by various researchers to serve specific purposes. For example viewing the object at different viewing angles and with different focal lengths within the Fresnel domain [50], reconstructing the image of a diffusively reflecting object using only the phase data of the complex amplitude [51], for digitally compensating the aberrations that arise during normal holographic reconstruction process [40] (by expanding the recorded wavefront to increase the spatial resolution), and to overcome the disruption in the visibility of the nearby focused objects due to the diffraction effects caused by the out of focus objects [52]. The relay lens used for increasing magnification works well at low magnification but causes aberration and noise at higher magnification.

In such a situation digital holographic microscopy [53] (DHM) is commonly adopted (**Figure 19**). This setup can successfully image liquid breakup elements as small as 5 μm. Only one beam is expanded with an objective lens and this serves both as a reference and an object beam. It is then passed directly through the liquid breakup region onto a CCD. The incident beam, in this case, is a coherent spherical wave emanating from a point source. This is achieved by passing the incident beam through a microscope objective followed by a pinhole. The pinhole size is of the order of the laser wavelength. A sample hologram of the liquid jet surface and its reconstructed image is shown in **Figure 20**. The resolution in DHM depends on (1) pin-hole size - this controls the spatial coherence and illumination cone; (2) numerical aperture - this is a function of the size and position of the CCD sensor; (3) pixel density and dynamic range - this controls fringe resolution and hologram noise level; and (4) wavelength of the laser light. Even though the depth of focus is small in this case owing to the large numerical aperture the digital reconstruction would allow one to focus at different depths of the object. Various modifications of DHM exist. In Digital Image Plane Holography (DIPH) [54], a plane of the fluid is illuminated with a laser sheet.

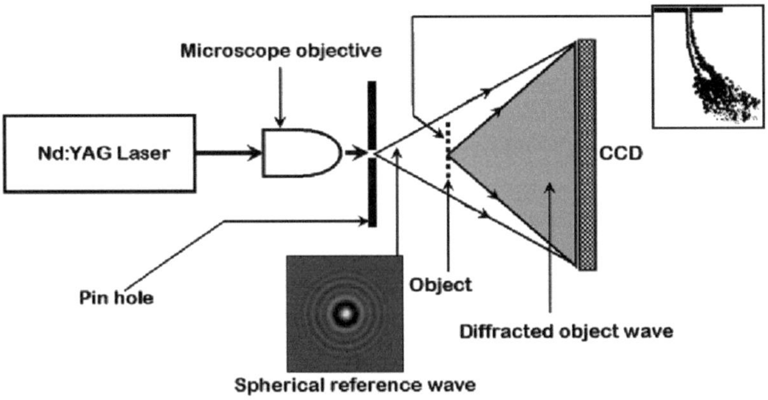

Figure 19.
Digital holographic microscopy setup.

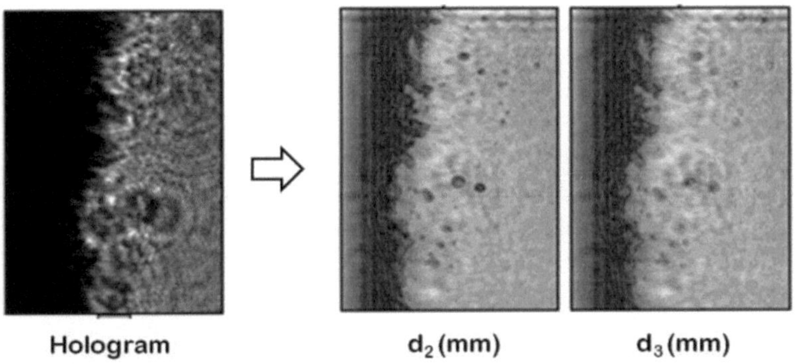

Figure 20.
Reconstructed holograms of a jet surface at different reconstruction distances.

The image of the illuminated fluid plane is then made to interfere with a reference beam on the CCD sensor which is imaged (**Figure 21**). The depth of focus is limited to the laser sheet thickness. Numerically reconstructing the holograms requires enormous processing power and memory. With Fast Fourier transform algorithms, the reconstructing time for a single hologram has been brought down to within a few tenths of a second on personal workstations. Another issue is adopting holography as a 3D particle characterization tool is its long depth of field [55]. Angular aperture in holography is limited to only a few degrees and this results in poor depth resolution. The finite pixel size of the image sensors limits the angle between the object and reference wave leading to the virtual and real images not being fully separated during hologram reconstruction. The field of view is also limited to the sensor size (or less if using a divergent beam).

The limited resolution capability of digital image sensors leads to the low spatial resolution of the hologram image. Some deconvolution methods based on the Weiner filter and iterative routines with point spread functions are used for overcoming the depth of focus limitation [56]. Other drawbacks include the reconstruction of spurious twin images of particles, leading to ghost images and multiple focusing around the actual depth location of the particles [57].

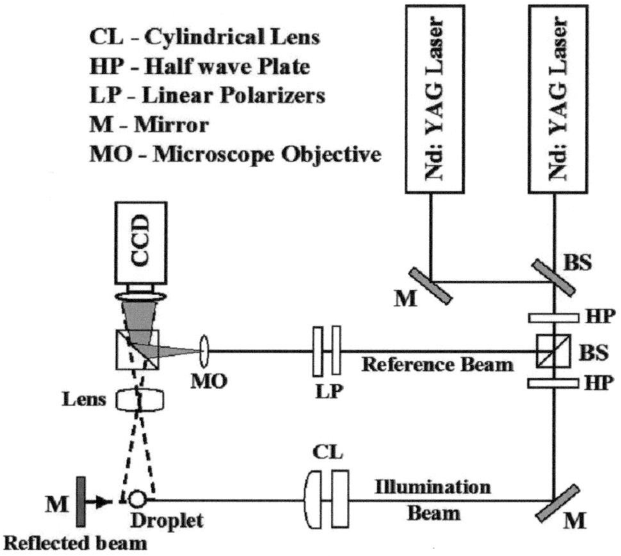

Figure 21.
Digital image plane holography (DIPH) setup [54].

3.4 X-ray phase contrast imaging

X-ray-based diagnostics have recently found increasing applicability in the study of liquid breakup due to some of its inherent advantages. The high penetrability of dense breakup regions and the ability to probe the internal mass distribution of breakup regions with a good time and spatial resolution makes it appealing. The x-ray beam experiences a significantly lower wave-shift as compared to visible light because of the large value of the index of refraction at optical wavelengths. This means the x-rays can be absorbed in specific wavelengths but they do not undergo appreciable scattering as in the case of visible light. Phase contrast imaging is based on utilizing the coherence properties of x-ray to enhance the edges of near-field Fresnel diffraction. Upon interaction with liquid jets, the x-rays suffer absorption along with a corresponding phase shift due to changes in phase media. Recording this phase shift on a CCD sensor enables us to distinguish between phase boundaries owing to improved image contrast and is termed as phase contrast imaging [58].

In a typical set-up, a synchrotron x-ray source is used whose undulator provides an extremely brilliant white monochromatic x-ray beam required for ultra-fast imaging purpose (**Figure 22**). The x-ray beam directly illuminates the breakup region but is tuned to the photon energy that does not get absorbed. When the x-rays pass through the sample they suffer attenuation and phase shift [59–62]. This creates small variations in the speed and direction of propagation of the x-rays. Interference effects are produced at the feature boundaries which have significant contrast. After interaction with the sample, the different wavefronts of the x-ray beam get diffracted by the sample and start to overlap and interfere with each other, giving rise to a contrasting pattern. This contrast depends on the Laplacian of the phase shift undergone by the beam upon its passage through the sample [63]. A fast scintillator crystal (LYSO:Ce or YAG:Ce) converts the x-ray phase contrast pattern into visible light which is then imaged into a CCD. Around the edges of features (ligaments and drops), an evolving pattern of light and dark fringes develop. This makes this technique particularly

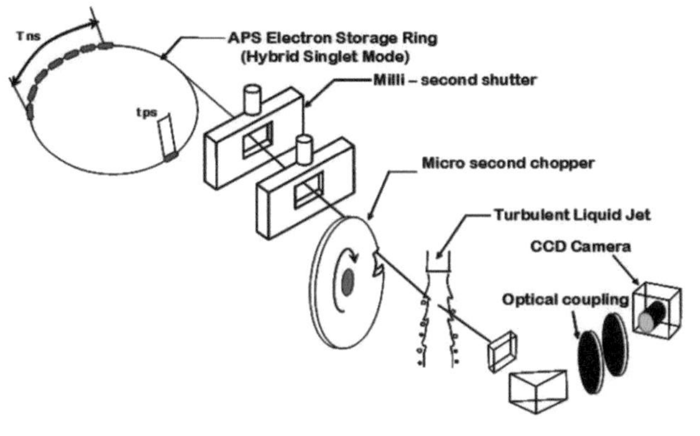

Figure 22.
X-ray phase contrast imaging experimental setup.

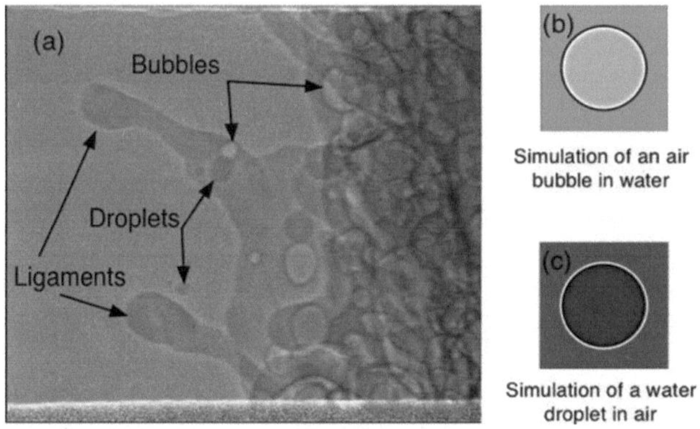

Figure 23.
(a) Phase contrast image of jet surface structures, and simulated images of (a) air bubble in water (c) water bubble in air respectively [59].

sensitive to boundaries and interfaces of different phase media [64]. Therefore a bubble (gas surrounded by liquid) will have a dark/bright outer edge while a droplet (liquid surrounded by gas) will have a bright/dark outer edge as illustrated in the simulations shown in **Figure 23**(b and c). The CCD camera is coupled to the scintillator using a microscope objective and a 45° mirror and records the images at a high speed. The exposure time (~150 ps) for each image is achieved by shuttering and timing the pulsed x-ray beam. In order to reduce the heat power in the beam, shuttering is usually employed whose synchronized operation could cut off more than 99% of the beam heat power.

The phase contrast images need to be normalized to exhibit the contrast/intensity gradients more efficiently. This technique is particularly suitable for studying dense primary breakup near-injector regions. The image resolution is generally a function of the detector resolution and the phase sensitivity, is a function of the source-specimen/specimen-detector distance. Phase contrast imaging has been applied to the study of spray breakup [58, 65] and as a simulation tool [66].

Some disadvantages of this technique are that the x-rays are difficult to manipulate and oftentimes the experiment needs to be designed in a way so as to be remotely controllable owing to safety issues. The breakup surface features e.g. ligaments and drops located at different axial distances are indistinguishable in the phase contrast x-ray images.

3.5 Ballistic imaging

Ballistic imaging [67–75] is a line-of-sight two-dimensional imaging technique that can overcome the effect of multiple scattering in dense liquid breakups. It can provide good quality images of the dense liquid breakup region which would otherwise be opaque to simpler imaging techniques. It derives its name from the term 'ballistic photons', the group of photons that do not suffer scattering when transmitted through highly turbid media because they travel the shortest path. This technique is based on a photon time of flight selection in which the transmitted photons are temporally selected, filtering out the multiply scattered photons which are time delayed depending on the optical length of the probed medium.

When light passes through a highly turbid medium, scattering occurs and the emerging beam comprises ballistic, snake, and diffuse photons (**Figure 24**).

3.5.1 Ballistic photons

These photons pass straight through the medium without scattering. They exit within the same solid angle that they entered and so travel the shortest path, and exit first.

3.5.2 Snake photons

These photons undergo negligible scattering, i.e. from a single scattering event to up to four scattering events and exit the medium along the same incident axis but with a larger solid angle than the ballistic photons.

3.5.3 Diffuse photons

These photons exit the medium after scattering multiple times (5 times or more). It has a large photon number density, is scattered into 4π steradian and exits last. They do not retain the memory of the structure within the material.

A typical Ballistic imaging setup [68], is shown in **Figure 25**. Ultra short light pulses (100 fs), isolated from all of the other background and noise photons (by using polarizer) are used as the imaging light to illuminate the break-up region. A time-resolved image is recorded using an ultra-fast time gate (full optical system such as the

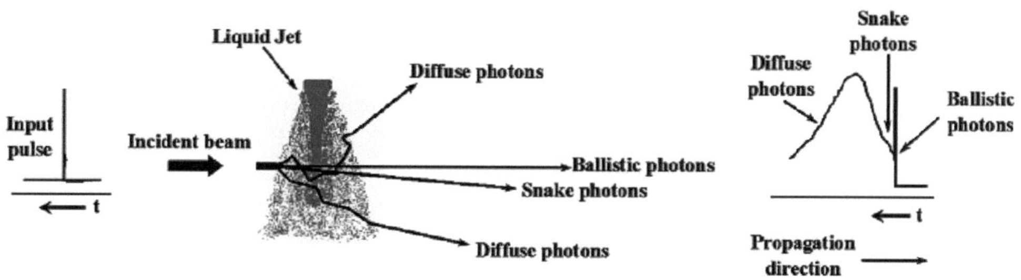

Figure 24.
Scattering through a dense media showing ballistic, snake, and diffuse photons [70].

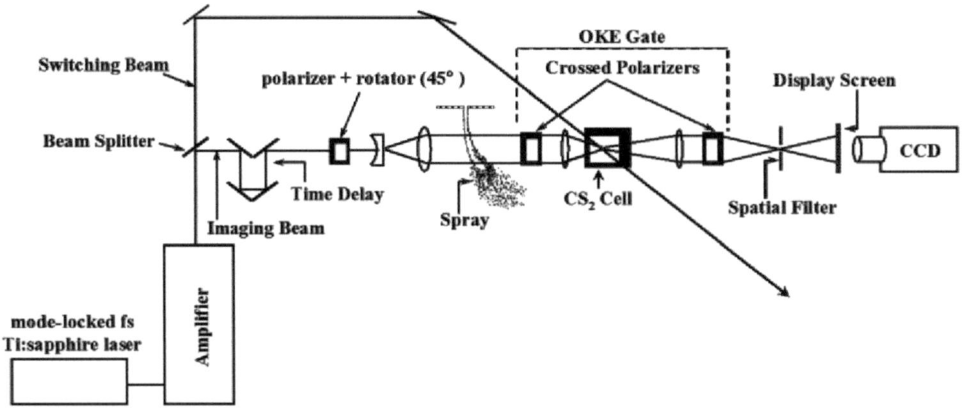

Figure 25.
Schematic of ballistic imaging setup.

optical Kerr gate), short enough to select photons with respect to their time of flight by separating them into ballistic, refractive, and scattered light which is then followed by constructing the image using just the selected class of photons.

The optical Kerr gate could be obtained by inducing a temporary birefringence in a Kerr active liquid between two crossed polarizers. The ballistic photons are attenuated by the large liquid structures inside a dense breakup region via absorption or refraction thus creating the intensity modulation required to produce an image.

These are used to form an undistorted, diffraction-limited image of structures inside the optically dense medium. The snake photons are also used along with the ballistic photons to improve the signal-to-noise ratio. Since highly turbid media leads to significantly more diffuse photons, they must be separated from the ballistic and snake photons. In **Figure 25**, a light source (1-kHz Ti-Sapphire regenerative amplifier, seeded with a Ti-Sapphire mode-locked laser oscillator) generates 150 femtosecond pulses centered around 800 nm). The light exiting the amplifier is linearly polarized and is split into an optical Kerr effect (OKE) switching beam and an imaging beam using a beam splitter. The OKE is a time gate consisting of a very fast shutter that selects only the leading edge of the image pulse containing ballistic and snake photons. To introduce the beams into the breakup region the linearly polarized imaging beam is made to pass through a polarizer followed by a rotation of the beam polarization by 45°. The imaging beam is then passed through a telescope which controls the beam size as it crosses the breakup region. The relay optics then focuses the beam through the OKE switch which serves as a shutter (2 ps) and is triggered by the switching pulses. When the switching pulse is absent image transfer to the display screen does not occur. The first polarizer in the OKE gate which is also the second polarizer used in the imaging beam, allows the polarization orientation of the imaging beam to pass through. The imaging beam is then focused into the Kerr active medium and up-collimated again. The second OKE polarizer is oriented normally with respected to the first one to block the unperturbed imaging beam. When the intense electric field of the switching pulse arrives at the Kerr active medium it rotates the polarization of the imaging beam, allowing most of it to pass through the second polarizer. After the imaging beam passes through the OKE gate, the image is relayed through a short-pass filter and displayed on a screen which is then imaged using a camera. By adjusting the length of the time delay segment of the imaging beam, it would be possible to

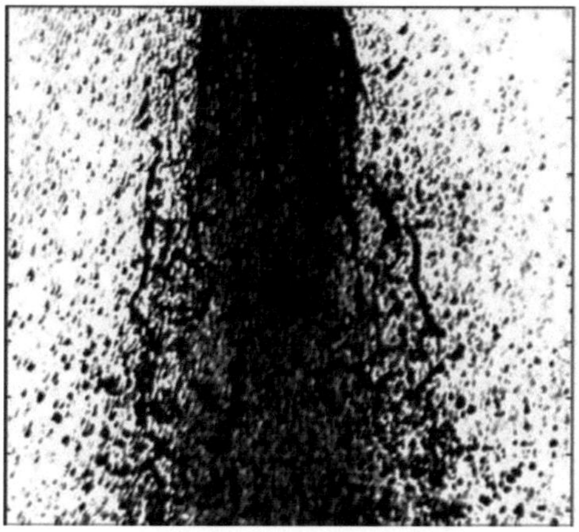

Figure 26.
Ballistic image of a jet breakup [2].

introduce a delay that would control the temporal overlap between the switching and imaging pulses when they arrive at the OKE gate for optimum time-gating.

The ballistic image thus obtained (**Figure 26**) reveals the main jet liquid column, droplet distribution, voids, and jet profile structures. The dark areas shown represent the liquid phase, the light areas are the gas phase, and the speckles and other spurious features are caused by diffraction.

If the droplets in the breakup region are smaller than the limit of spatial resolution for the setup (~40–50 μm) they go undetected which is a disadvantage. The spatial resolution is limited by the CS_2 liquid contained in the CS_2 cell located in the Fourier plane which acts as a spatial filter to reduce the scattering noise.

3.6 Liquid jet Core illumination

Optical connectivity or liquid jet core illumination is another relatively new technique for studying atomization. Here, a light guide illuminates the liquid jet from within the injecting nozzle in a direction parallel to the flow [76, 77].

Here the liquid jet acts as an optical fiber through which the light propagates and is interrupted only at the breakup region (**Figure 27**). The continuous portion of the liquid jet is made to fluoresce by adding a fluorescing dye (Rhodamine WT dye). The laser beam is steered by means of a light-guide tube or an optical fiber through the liquid jet nozzle. The fluorescing dye in the liquid emits fluorescent light along the laser beam path and is interrupted at the breakup region. The intensity of the beam should be sufficient to identify this position. The advantage of introducing the laser beam from within the liquid nozzle is that it is minimally attenuated and multiple scattered and maintains a low light intensity beyond the surface of the liquid jet. The addition of the dye aids in the emission of longer wavelengths of fluorescing light which can be detected separately from the incoming laser beam thus avoiding background noise on images due to the scattered light. The fluorescing light is emitted from everywhere inside the liquid core, associated with the volume of the liquid. The liquid jet surface features act as minute mirrors to focus the laser beam which produces bright spots along the liquid surface and could be identified as locations for significant changes in the surface structure. Short-duration laser pulses allow the

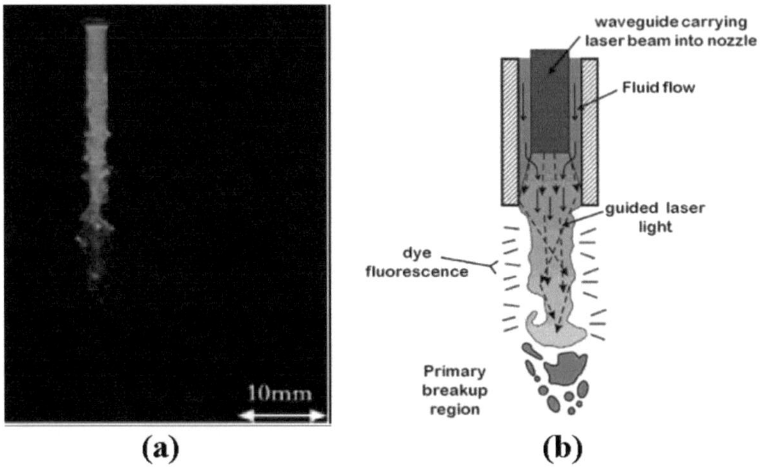

Figure 27.
(a) Liquid jet core illumination [76] (b) principle of optical connectivity [2].

image to be frozen in time. The imaging system is a CCD camera fitted with a low-pass filter to suppress the scattered light from the laser beam.

This diagnostic promises improved measurements for atomization occurring at the nozzle exit since scattering (caused by the diffusion of the laser light by the droplets) persists further downstream of the breakup region. Light propagation in narrow ligaments connecting the detaching masses of liquid and the surrounding droplets causes the light to become diffuse and its intensity to be significantly reduced after the core breakup region. The detached droplets do not fluoresce and the area surrounding the breakup point does not illuminate.

As the jet becomes increasingly turbulent, the transition of the smooth jet surface to a rough and wavy one results in the angle of incidence between the laser light rays and the jet surface becoming less than the critical angle for total internal reflection. This leads to increased scattering, loss of fluorescence intensity, and bright spots created due to the focusing of laser light by jet surface features which act as minute mirrors. In certain situations like at high Reynolds numbers, the liquid core can suddenly change thickness drastically, become very thin, and then thick again. The fluid constriction thus created can cause significant loss of laser light. Moreover, it does not work for a situation requiring droplet size measurements or location determination, or their velocity measurements as is often possible with other techniques.

3.7 Mie scattering

Mie scattering is an elastic scattering that occurs when the size of the particles in the liquid breakup is of the order of the scattered wavelengths. Mie-scattering intensity is proportional to the total liquid surface area. A number of researchers have used Mie imaging for investigating fuel sprays [78–84]. An instantaneous Mie scattering image of a jet in crossflow [78] is shown in **Figure 28**, where the inset represents the original Mie image which is then inverted to yield a dark image of the jet against a white background.

A Mie scattering setup for two-dimensional spray visualization [82–86] is shown in **Figure 29**. A wide planar laser sheet is formed using a cylindrical lens, made to pass

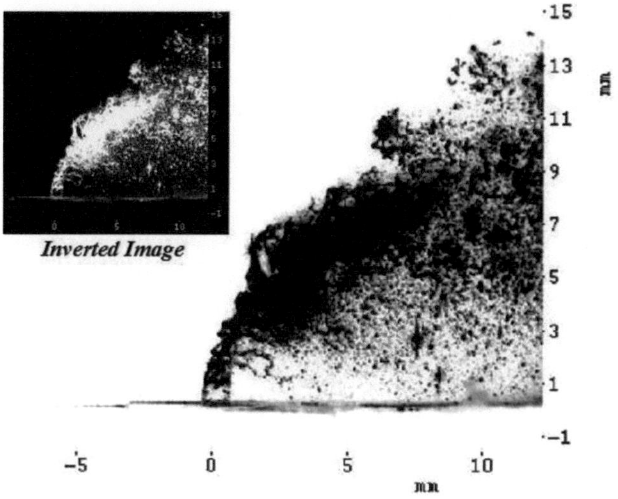

Figure 28.
Instantaneous Mie scattering image [78].

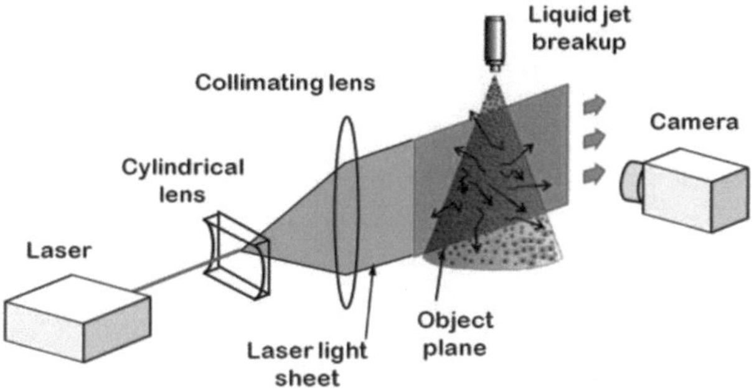

Figure 29.
Mie scattering setup for two-dimensional spray visualization.

through a particular plane of the liquid jet. The scattered signal normal to the illuminated plane is imaged onto an ICCD camera fitted with a narrowband interference filter. The images are averaged for noise removal followed by background subtraction, uniform field correction, and intensity normalization. The clarity of the Mie image depends on the extent of multiple scattering. When the droplet concentration is low and the breakup dimension is small, the detected signal does not suffer attenuation. However, when droplet concentration is high the incident beam becomes attenuated as it travels the breakup region, 'secondary scattering' from droplets lying between the incident beam and the detector and 'multiply scattering' by the droplets in the surrounding atomized region result in "extraneous light" to be detected. Since the Mie intensity is proportional to the square of the particle diameter ($\sim d_p^2$) a CCD sensor having a large dynamic range is required to simultaneously image the large liquid core and the smallest droplets. Because the scattered cross section is always larger than the corresponding ligament/droplet diameter, it is difficult to carry out accurate size measurements and distributions from the Mie signals. Spatial resolution is also a significant limitation of Mie imaging.

Comparisons are often made between shadowgraphy and Mie imaging. However, while Mie imaging is the visualization of elastically scattered light normal to the plane of the incident light sheet passing through the breakup region, shadowgraphy is the image of the shadow of the breakup region formed by a collimated beam of light.

3.8 Planar laser-induced fluorescence

Planar laser-induced fluorescence (PLIF) is a technique [87–93] where a plane laser light sheet is incident on a cross-section of the fluid mixed with a fluorescent dye. The dye causes the liquid plane to fluoresce and this fluoresced pattern is then imaged by a camera. A part of the laser excitation lies within the absorption spectrum of the dye. **Figure 30** shows a typical PLIF setup.

In the fluorescence linear regime, the fluorescence intensity is proportional to the laser excitation intensity. The dye has a large separation between the absorption and emission spectra (e.g. Rhodamine WT/6G/B, and Fluorescein). A narrow-band optical filter allows only the fluorescence wavelengths to be imaged by the camera. **Figure 31** shows a PLIF image of a jet primary breakup. The characteristic decay time (time necessary for 86.5% of fluorescence intensity to be emitted) and fluorescence intensity require calibration [20, 89]. The PLIF technique suffers from a non-uniform

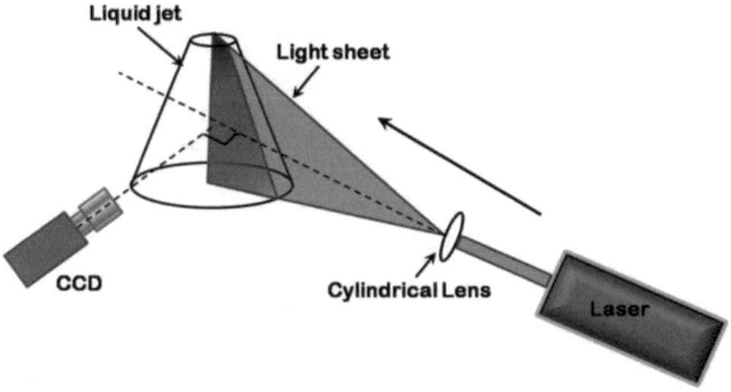

Figure 30.
PLIF setup.

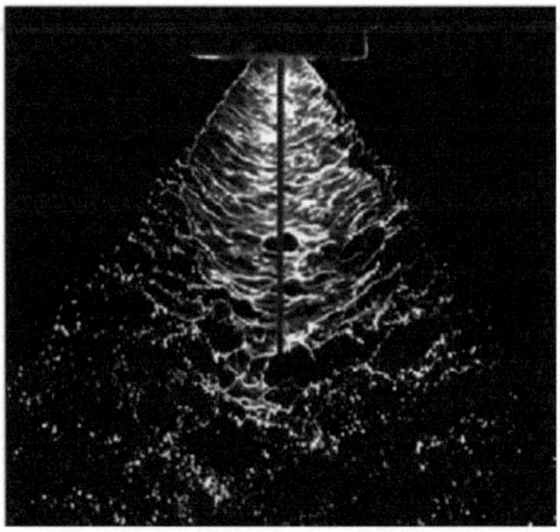

Figure 31.
PLIF image showing the liquid jet sheet breakup [88].

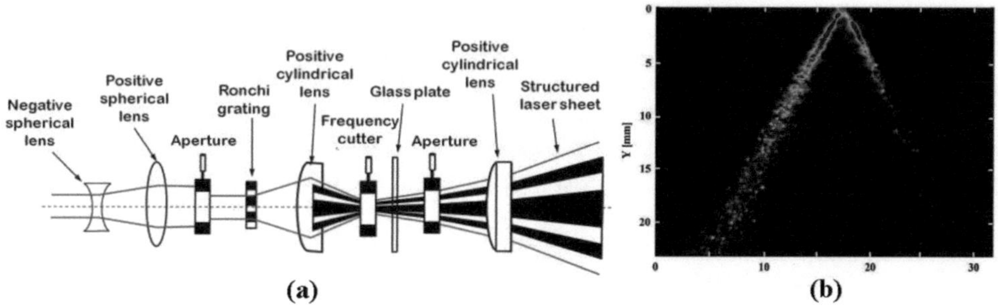

Figure 32.
(a) SLIPI beam setup [2] (b) SLIPI Mie scattering planar image of hollow cone spray [2].

intensity profile of the planar laser sheet, and light intensity reduction due to absorption by the background dye concentration. High dye concentrations produce non-linearities in fluorescence because of absorption-related changes in the excitation intensity. The camera exposure time depends on the spectral sensitivity of the camera's sensor at the fluorescing wavelength.

Light emitted from the object plane in PLIF could be either Mie scattered radiation or fluorescence light. To reduce the scattering, the Structured Laser Illumination Planar Imaging (SLIPI) technique is carried out (**Figure 32**a). SLIPI [94–96] generates a recognizable signature in the incident planar beam (by time-based modulation of the excitation source) that yields signal photons. The multiply scattered photons are discarded due to their lack of this signature. **Figure 32**(b) shows a SLIPI Mie scattering planar image in a hollow cone spray.

3.9 Thermal imaging

Thermal imaging is used for temperature visualization in the breakup regime. It uses planar laser-induced emission (fluorescence/phosphorescence) of a suitable excitation wavelength in the UV region and thermographic phosphors or organic tracer molecules. The two primary methods under thermal imaging are thermographic fluorescence and thermographic phosphorescence.

3.9.1 Thermographic fluorescence

In this technique [7, 97–100], a fluorescent additive is added to the hydrocarbon liquid in the form of a monomer (M), and an exciplex forming an organic molecule called quencher (N_0). When this sample is irradiated, M is electronically excited to form M* which then reacts with N_0, which is in the ground state to form the organic exciplex (E*) which is an excited state complex. The reaction is represented by

$$M^* + N_0 \Leftrightarrow E^* \quad (1)$$

Fluorescence is observed from both M* and E*. Typical fluorescence lifetimes are less than 100 nanoseconds and the populations of M* and E* are strongly temperature dependent. By comparing the decay lifetimes or the intensity ratio between two or several emission lines to a calibrated standard, the liquid breakup droplet temperature distribution can be obtained. A typical experimental setup resembles the PLIF setup [97]. An Nd:YAG laser source produces a light sheet which is incident on a liquid plane to excite the monomer/quencher pairs leading to exciplex formation and fluorescence from the breakup region. This is photographed at right angles to the plane of the laser light pulse. The liquid and vapor phase images are imaged with a single ICCD camera using UV objective and a stereoscope. Different emission lines correspond to different temperature ranges. The spatial profile of the breakup, droplets, and vapor are distinguishable and measurable. For a multi-component fuel mixture, a two-color laser-induced fluorescence technique can be adopted [101, 102] in which the additive (pyrromethene) exhibits a highly different temperature-sensitive fluorescence spectrum for two spectral bands of detection. Rhodamine B is often used as a fluorescent temperature sensor as it exhibits significant temperature dependence. This technique can measure the thermal transport properties (mean and fluctuating characteristics of the dynamic and thermal fields) of a turbulent heated jet undergoing atomization.

3.9.2 Thermographic Phosphorescence

Thermographic phosphors (e.g. La_2O_2S:Eu or Mg_4FGeO_6:Mn) are temperature-sensitive materials that when excited by UV light exhibit temperature-dependent emission characteristics. Thermographic phosphors are mixed with the liquid forming the jet. A thin laser sheet is then directed at the liquid plane to be imaged. This causes phosphor excitation and subsequent emission from the breakup region [103, 104]. After excitation, the lifetime of a particular emission line is monitored to obtain temperature measurements.

The emitted light is passed through an interference filter, and recorded by a fast-framing ICCD camera. After image acquisition, the pixel positions of the ICCD detectors are correlated with one another. Using an exponential fitting procedure, the lifetime and temperature information is then extracted. A two-dimensional temperature image is thus obtained, as shown in **Figure 33** It shows a water spray temperature distribution image. Since the temperature of the droplets of different sizes after jet breakup changes due to evaporation and convection in comparison to the main jet core, their detection becomes possible using this method.

This technique has high accuracy; the advantage of remote detection and a high signal yield; and is independent of pressure variations. The excitation energies for both fluorescence and phosphorescence should be below a threshold value as high excitation energies lead to luminescence saturation. Fluorescence lifetime is on the order of nanoseconds whereas phosphorescence lifetime ranges between the orders of microseconds to milliseconds. The decay time is often a very sensitive function of

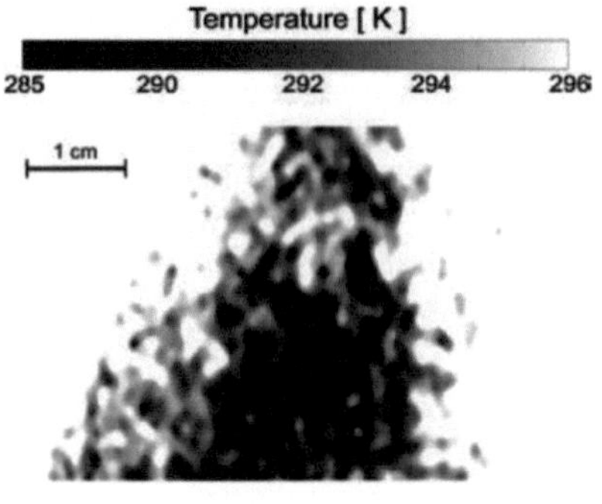

Figure 33.
Single shot water spray image using thermographic phosphors [103].

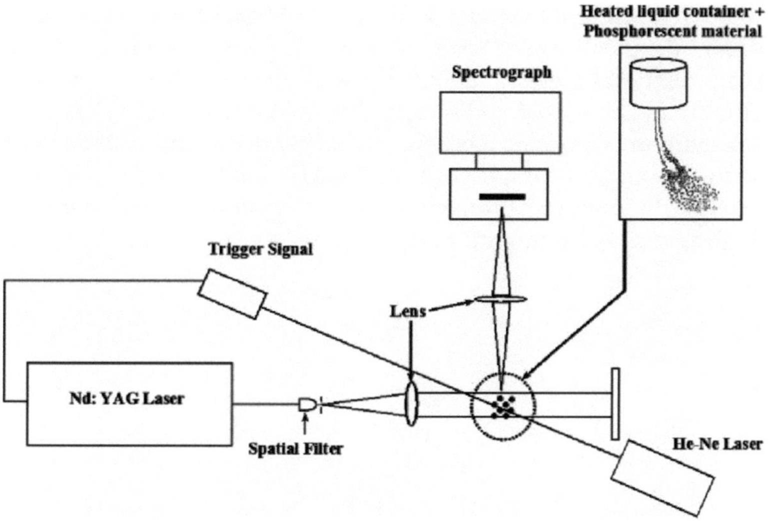

Figure 34.
Experimental setup of thermal imaging by the spectral method.

temperature [105]. So any phenomenon is limited to being observed if only it occurs within the excited state period of the phosphor and the speed is essentially dependent on the camera frame rate (MHz). Combining both the fluorescence and phosphorescence techniques has shown to be successful [106, 107] in understanding the breakup and mixing of liquid fuels at high pressure and temperature conditions similar to those in combustion chambers. **Figure 34** shows the experimental setup of thermal imaging by the spectral method.

4. Conclusion

The need to develop newer spray characterization techniques has always existed due to an increasing emphasis on understanding the different unsteady processes involved in liquid atomization, primarily for the aerodynamics and combustion industry. This involves complete characterization of flow fields, in terms of density, temperature, pressure, and flow velocity. Among the techniques discussed here, four are transillumination techniques (shadowgraphy, holography, x-ray phase contrast, ballistic) where an incident beam is passed through the liquid breakup and imaged on the other side, and one is an internal illumination technique (jet core illumination), one is a scattering (Mie) technique, and the rest are planar laser fluorescence/phosphorescence technique. Besides these high-speed ordinary imaging is based on reflection. From a study of the experimental diagnostics that are presently pursued by the scientific community, we can observe that even though the experimental techniques have remarkably improved over time in providing us with more information about the breakup process they are still limited in their applicability as far as meeting the complete demands is concerned. Some of the alternative non-intrusive laser-based diagnostic techniques that exist at presents such as Phase Doppler Particle Analyzer (PDPA), Phase Doppler Anemometry (PDA), Laser Doppler Anemometry (LDA), Laser Diffraction particle sizing, Laser-Induced Incandescence (LII), and Chemiluminescence Imaging, work mostly in the gaseous/vapor phase [108] or are mainly concerned with determining the particle sizes, velocities, and volume fraction. Almost all of the present diagnostics are restricted to low Reynolds number flows to yield accurate Spatio-temporal measurements. For some very small length and time scale fluid flow situations, the spatial and temporal resolution, as well as frequency, is beyond the resolution of present-day high-speed video cameras. There is a technological need for developing ultra-high-speed cameras of sufficient dynamic range/resolution and even techniques to capture the high-speed atomization process with extremely high spatial and temporal resolution.

Author details

Anu Osta
Rowan University, Glassboro, USA

*Address all correspondence to: anu.osta@hotmail.com

References

[1] Coghe A, Cossali GE. Quantitative optical techniques for dense sprays investigation: A survey. Optics and Lasers in Engineering. 2012;**50**:46-56

[2] Linne M. Imaging in the optically dense regions of a spray: A review of developing techniques. Progress in Energy and Combustion Science. 2013; **39**:403-440

[3] Fansler TD, Parrish SE. Spray measurement technology: A review. Measurement Science and Technology. 2015;**26**(012002):34

[4] Soid SN, Zainal ZA. Spray and combustion characterization for internal combustion engines using optical measuring techniques - a review. Energy. 2011;**36**:724-741

[5] Sridhara SN, Raghunandan BN. Photographic investigations of jet disintegration in Airblast sprays. Journal of Applied Fluid Mechanism. 2010;**3**(2): 111-123

[6] Parker T, Rainaldi LR, Rawlins WT. A comparative study of room-temperature and combusting fuel sprays near the injector tip using infrared laser diagnostics. Atomization and Sprays. 1998;**8**(5):565-600

[7] Khalid AH, Kontis K. Thermographic phosphors for high temperature measurements: Principles. Current State of the Art and Recent Applications, Sensors. 2008;**8**:5673-5744

[8] Thoroddsen ST, Etoh TG, Takehara K. High-speed imaging of drops and bubbles. Annual Review of Fluid Mechanics. 2008;**40**:257-285

[9] Bachalo WD. Spray diagnostics for the 21st century. Atomization and Sprays. 2000;**10**:439-474

[10] Vu Truong Son D, Goji Etoh T, Tanaka M, Hoang Dung N, Le Cuong V, Takehara K, et al. Toward 100 mega-frames per second: Design of an Ultimate Ultra-High-Speed Image Sensor. Sensors. 2010;**10**:16-35

[11] Crua C, Shoba T, Heikal M, Gold M, Higham C. High-speed microscopic imaging of the initial stage of diesel spray formation and primary breakup. SAE International. 2010;**01**:2247

[12] Lasheras JC, Villermaux E, Hopfinger EJ. Break-up and atomization of a round water jet by a high-speed annular air jet. Journal of Fluid Mechanics. 1998;**357**(1):351-379

[13] Zhu Y, Oguz HN, Prosperetti A. On the mechanism of air entrainment by liquid jets at a free surface. Journal of Fluid Mechanics. 2000;**404**: 151-177

[14] Varga CM, Lasheras JC, Hopfinger EJ. Initial breakup of a small-diameter liquid jet by a high-speed gas stream. Journal of Fluid Mechanics. 2003;**497**:405-434

[15] Hoyt JW, Taylor JJ. Turbulence structure in a water jet discharging in air. Physics of Fluids. 1977;**20**: S253-S257

[16] Ben-Eliezer E, Marom E, Konforti N, Zalevsky Z. Experimental realization of an imaging system with an extended depth of field. Applied Optics. 2005;**44**: 2792-2798

[17] Bauer D, Chaves H, Brucker C. A multiple-segment long-distance microscope for flow visualization and measurements. Measurement Science and Technology. 2009;**20**:1-4

[18] Nagahara H, Kuthirummal S, Zhou C, Nayar S. Flexible depth of field photography. Proceedings of European Conference on Computer Vision. 2008; **4**:60-73

[19] Dowski ER, Cathey WT. Extended depth of field through wave-front coding. Journal of Applied Optics. 1995; **34**(11):1859-1866

[20] Leong MY, McDonell VG, Samuelsen GS. Effect of ambient pressure on an airblast spray injected into a crossflow. Journal of Propulsion and Power. 2001;**17**(5)

[21] Castrejon-Garcia R, Castrejon-Pita JR, Martin GD, Hutchings IM. The shadowgraph imaging technique and its modern application to fluid jets and drops. Revista Mexicana de Fisica. 2011; **57**:266-275

[22] Osta AR, Sallam K. Nozzle-geometry effects on upwind-surface properties of turbulent liquid jets in gaseous crossflow. Journal of Propulsion and Power. 2010;**26**(5):936-946

[23] Timmerman BH, Bryanston-Cross PJ, Skeen AJ, Tucker PG, Jefferson-Loveday RJ, Dunkley P, et al. High-speed digital Shadowgraphy for high frequency shock tracking in supersonic flows. In: 7th Symposium on Measuring Techniques in Transonic and Supersonic Flows in Cascades and Turbomachines. Sweden: Stockholm; 2004

[24] Sinha A, Prakash RS, Madan Mohan A, Ravikrishna RV. Airblast spray in crossflow – Structure, trajectory and droplet sizing. International Journal of Multiphase Flow. 2015;**72**:97-111

[25] Bougie B, Tulej M, Dreier T, Dam NJ, Meulen JJ, Gerber T. Optical diagnostics of diesel spray injections and combustion in a high-pressure high-temperature cell. Applied Physics B. 2005;**80**(8): 1039-1045

[26] Lee K, Aalburg C, Diez FJ, Faeth GM, Sallam KA. Primary breakup of turbulent round liquid jets in uniform crossflows. AIAA Journal. 2007;**45**(8): 1907-1916

[27] Klein-Douwel RJH, Frijters PJM, Somers LMT, de Boer WA, Baert RSG. Macroscopic diesel fuel spray shadowgraphy using high speed digital imaging in a high pressure cell. Fuel. 2007;**86**:1994-2007

[28] Watanabe H, Okazaki K. Visualization of secondary atomization in emulsified-fuel spray flow by shadow imaging. Proceedings of the Combustion Institute. 2013;**34**:1651-1658

[29] Trainoff SP, Cannell DS. Physical optics treatment of the shadowgraph. Physics of Fluids. 2002;**14**(4):1340-1363

[30] Wu PK, Kirkendall KA, Fuller RP, Nejad AS. Breakup processes of liquid jets in subsonic crossflows. Journal of Propulsion and Power. 1997;**13**(1):64-73

[31] Sallam K, Aalburg C, Faeth GM. Breakup of round nonturbulent liquid jets in gaseous crossflow. AIAA Journal. 2004;**42**(12):2529-2540

[32] Settles GS. Schlieren and Shadowgraph Techniques: Visualizing Phenomena in Transparent Media. Berlin, Heidelberg, Germany: Springer; 2001. p. 376. Available from: https://doi.org/10.1007/978-3-642-56640-0

[33] Muirhead JC, McCallum FL. Improvement in spark shadowgraph technique. The Review of Scientific Instruments. 1959;**30**:830-831

[34] Fuller RP, Wu PK, Kirkendall KA, Nejad AS. Effects of injection angle on

atomization of liquid jets in transverse airflow. AIAA Journal. 2000;**38**(1):64-72

[35] Biss MM, Settles GS, Hargather MJ, Dodson LJ, Miller JD. High-speed digital Shadowgraphy of shock waves from explosions and gunshots. Shock Waves. 2009;**II**:91-96

[36] Volpe JA, Settles GS. Laser-induced gas breakdown as a light source for Schlieren and shadowgraph particle image velocimetry. Optical Engineering. 2006;**45**(8):080509

[37] Klinner J, Willert C. Tomographic Shadowgraphy for three-dimensional reconstruction of instantaneous spray distributions. Experiments in Fluids. 2012;**53**:531-543

[38] Wang Q, Zhang Y. High speed stereoscopic shadowgraph imaging and its digital 3d reconstruction. Measurement Science and Technology. 2011;**22**(6):1-9

[39] Gabor D. Holography, 1948-1971. Science, New Series. 1972;**177**(4046): 299-313

[40] Schnars U, Juptner W. Digital recording and numerical reconstruction of holograms. Measurement Science and Technology. 2002;**13**:R85-R101

[41] Frauel Y, Naughton TJ, Matoba O, Tajahuerce E, Javidi B. Three-dimensional imaging and processing using computational holographic imaging. Proceedings of the IEEE. 2006; **94**(3):636-653

[42] Leith EN, Upatnieks J. Holograms - their properties and uses (hologram photography, principles, techniques and application). SPIE Journal. 1965;**4**:3-6

[43] Kubota T. 48 years with holography. Optical Review. 2014;**21**(6):883-892

[44] Milgram JH, Li W. Computational reconstruction of images from holograms. Applied Optics. 2002;**41**: 853-864

[45] Muller J, Kebbel V, Juptner W. Characterization of spatial particle distributions in a spray-forming process using digital holography. Measurement Science and Technology. 2004;**15**: 706-710

[46] Santangelo PJ, Sojka PE. A holographic investigation of the near-nozzle structure of an effervescent atomizer-produced spray. Atomization and Sprays. 1995;**5**(2):137-155

[47] Burke J, Hess CF, Kebbel V. Digital holography for instantaneous spray diagnostics on a plane. Particle and Particle Systems Characterization. 2003; **20**:183-192

[48] Miller B, Sallam KA, Bingabr M, Lin K-C, Carter CD. Breakup of aerated liquid jets in subsonic crossflow. Journal of Propulsion and Power. 2008;**24**(2): 253-258

[49] Lee J, Miller B, Sallam KA. Demonstration of digital holographic diagnostics for the breakup of liquid jets using a commercial-grade CCD sensor. Atomization and Sprays. 2009;**19**(5): 445-456

[50] Yu L, An Y, Cai L. Numerical reconstruction of digital holograms with variable viewing angles. Optics Express. 2002;**22**(10):1250-1257

[51] Yamaguchi I, Yamamoto K, Mills GA, Yokota M. Image reconstruction only by phase data in phase-shifting digital holography. Applied Optics. 2006;**45**:975-983

[52] Monnom O, Dubois F, Yourassowsky C, Legros JC.

Improvement in visibility of an In-focus reconstructed image in digital holography by reduction of the influence of out-of-focus objects. Applied Optics. 2005;**44**(18):3827-3832

[53] Garcia-Sucerquia J, Xu W, Jericho SK, Klages P, Jericho MH, Kreuzer HJ. Digital In-line holographic microscopy. Applied Optics. 2006;**45**: 836-850

[54] Palero V, Arroyo MP, Soria J. Digital holography for micro-droplet diagnostics. Experiments in Fluids. 2007;**43**:185-195

[55] Choo YJ, Kang BS. The characteristics of the particle position along an optical Axis in particle holography. Measurement Science and Technology. 2006;**17**:761-770

[56] Latychevskaia T, Gehri F, Fink HW. Depth-resolved holographic reconstructions by three-dimensional deconvolution. Optics Express. 2010; **18**(21):22527-22544

[57] Gire J, Denis L, Fournier C, Thiebaut E, Soulez F, Ducottet C. Digital holography of particles: Benefits of the 'Inverse Problem' approach. Measurement Science and Technology. 2008;**19**(7):074005

[58] Wang YJ, Im K, Fezzaa K, Lee WK, Wang J. Quantitative x-ray phase-contrast imaging of air-assisted water sprays with high Weber numbers. Applied Physics Letters. 2006;**89**:151913

[59] Osta AR, Lee J, Sallam KA, Fezzaa K. Study of the effects of the injector length/diameter ratio on the surface properties of turbulent liquid jets in still air using X-ray imaging. International Journal of Multiphase Flow. 2012;**38**(1):87-98

[60] James RW. The Optical Principles of the Diffraction of X-Rays. Ithaca, NY, USA: Cornell University Press; 1965

[61] El-Ghazaly M, Backe H, Lauth W, Kube G, Kunz P, Sharafutdinov A, et al. X-Ray Phase Contrast Imaging at MAMI. European Physical Journal A: Hadrons and Nuclei. 2006;**28**(S01):197-208

[62] Powell CF, Yue Y, Poola R, Wang J. Time-resolved measurements of supersonic fuel sprays using synchrotron X-rays. Journal of Synchrotron Radiation. 2000;**7**:356-360

[63] Guigay JP, Langer M, Boistel R, Cloetens P. Mixed transfer function and transport of intensity approach for phase retrieval in the Fresnel region. Optics Letters. 2007;**32**:1617-1619

[64] Wilkins SW, Gureyev TE, Gao D, Pogany A, Stevenson AW. Phase-contrast imaging using polychromatic hard X-rays. Nature. 1996;**384**(6607): 335-338

[65] Moon S, Liu Z, Gao J, Dufresne E, Fezzaa K, Wang J. Ultrafast X-ray phase-contrast imaging of high-speed fuel sprays from a two-hole diesel nozzle. In: 22nd Annual Conference on Liquid Atomization and Spray Systems. Cincinnati, OH: ILASS; 2010

[66] Peterzol A, Berthier J, Duvauchelle P, Ferrero C, Babot D. X-ray phase contrast image simulation. Nuclear Instruments and Methods in Physics Research Section B: Beam Interactions with Materials and Atoms. 2007;**254**(2):307-318

[67] Sedarsky D, Paciaroni M, Berrocal E, Petterson P, Zelina J, Gord J, et al. Model validation image data for breakup of a liquid jet in crossflow: Part I. Experiments in Fluids. 2010;**49**(2): 391-408

[68] Sedarsky D, Gord J, Carter C, Meyer T, Linne M. Fast-framing ballistic

imaging of velocity in an aerated spray. Optics Letters. 2009;**34**:2748-2750

[69] Linne MA, Paciaroni M, Hall T, Parker T. Ballistic imaging of the near field in a diesel spray. Experiments in Fluids. 2006;**40**(6):836-846

[70] Linne MA, Paciaroni M, Berrocal E, Sedarsky D. Ballistic imaging of liquid breakup processes in dense sprays. Proceedings of the Combustion Institute. 2009;**32**(II):2147-2161

[71] Paciaroni M, Linne M. Single-shot, two-dimensional ballistic imaging through scattering media. Applied Optics. 2004;**43**:5100-5109

[72] Paciaroni M, Hall T, Delplanque JP, Parker T, Linne M. Single-shot two-dimensional ballistic imaging of the liquid Core in an atomizing spray. Atomization and Sprays. 2006;**16**(1): 51-70

[73] Idlahcen S, Rozé C, Méès L, Girasole T, Blaisot JB. Sub-picosecond ballistic imaging of a liquid jet. Experiments in Fluids. 2012;**52**:289-298

[74] Sedarsky D, Paciaroni M, Linne M, Gord J, Meyer T. Velocity imaging for the liquid–gas interface in the near field of an atomizing spray: Proof of concept. Optics Letters. 2006;**31**:906-908

[75] Idlahcen S, Mees L, Roze C, Girasole T, Blaisot J. Time gate, optical layout, and wavelength effects on ballistic imaging. Journal of the Optical Society of America. A. 2009;**26**(9): 1995-2004

[76] Charalampous G, Hardalupas Y, Taylor A. Novel technique for measurements of continuous liquid jet Core in an atomizer. AIAA Journal. 2009;**47**(11):2605-2615

[77] Charalampous G, Hadjiyiannis C, Hardalupas Y. Proper orthogonal decomposition analysis of photographic and optical connectivity time resolved images of an atomizing liquid jet. In: 24th Ann. Conf. Liquid Atom. Spray Sys. Portugal: ILASS; 2011. pp. 1-8

[78] Elshamy OM, Tambe SB, Cai J, Jeng SM. Structure of liquid jets in subsonic crossflow at elevated ambient pressures. In: 44th AIAA Aerospace Sciences Meeting and Exhibit. Reno: AIAA; 2006. pp. 1-11

[79] Smallwood GJ, Gulder OL, Snelling DR. The structure of the dense Core region in transient diesel sprays. 25th Symposium (International) on Combustion. 1994;**25**(1):371-379

[80] Fansler TD, Drake MC. "Designer diagnostics" for developing direct-injection gasoline engines. Journal of Physics Conference Series. 2006;**45**:1-17

[81] Sick V, Drake MC, Fansler TD. High-speed imaging for direct-injection gasoline engine Research and Development. Experiments in Fluids. 2010;**49**(4):937-947

[82] Stenzler JN, Lee JG, Santavicca DA. Penetration of liquid jets in a crossflow. Atomization and Sprays. 2006;**16**(8): 887-906

[83] Tambe S, Elshamy O, Jeng S. Spray properties of liquid jets injected transversely into a shear layer. In: 43rd AIAA/ASME/SAE/ASEE Joint Propulsion Conference & Exhibit. Cincinnati, OH: AIAA; 2007. pp. 2007-5695

[84] Rachner M, Becker J, Hassa C, Doerr T. Modelling of the atomization of a plain liquid fuel jet in crossflow at gas turbine conditions. Aerospace Science and Technology. 2002;**6**(7):495-506

[85] Han D, Mungal MG, Orozco V. Gross-entrainment behavior of turbulent jets injected obliquely into a uniform crossflow. AIAA Journal. 2000;**38**(9): 1643-1649

[86] Siebers, D.L., (1998) Liquid-Phase Fuel Penetration in Diesel Sprays, International Congress & Exposition, Detroit, MI, Paper No. 980809.

[87] Niederhaus CE, Champagne FH, Jacobs JW. Scalar transport in a swirling transverse jet. AIAA Journal. 1997; **35**(11):1697-1704

[88] Cloeter MD, Qin K, Patil P, Smith B. Planar laser induced fluorescence (PLIF) flow visualization applied to agricultural spray nozzles with sheet disintegration; influence of an oil-in-water emulsion. In: 22nd Annual Conference on Liquid Atomization and Spray Systems. Cincinnati, OH: ILASS; 2010

[89] Pastor JV, Lopez JJ, Julia JE, Benajes JV. Planar laser-induced fluorescence fuel concentration measurements in isothermal diesel sprays. Optics Express. 2002;**10**:309-323

[90] Davidson MJ, Pun KL. Weakly Advected jets in Cross-flow. Journal of Hydraulic Engineering. 1999;**125**(1): 47-58

[91] Wieske P, Wissel S, Grünefeld G, Pischinger S. Improvement of LIEF by wavelength-resolved acquisition of multiple images using a single CCD detector– Simultaneous 2D measurement of air/fuel ratio, temperature distribution of the liquid phase and qualitative distribution of the liquid phase with the multi-2D technique. Applied Physics B. 2006; **83**(2):323-329

[92] Webster DR, Longmire EK. Jet pinch-off and drop formation in immiscible liquid–liquid systems. Experiments in Fluids. 2001;**30**(1):47-56

[93] Milosevic IN, Longmire EK. Pinch-off modes and satellite formation in liquid/liquid jet systems, Int. Journal of Multiphase Flow. 2002;**28**:1853-1869

[94] Kristensson E, Berrocal E, Ritcher M, Pettersson S-G, Aldén M. High-speed structured planar laser illumination for contrast improvement of two-phase flow images. Optics Letters. 2008;**33**(23):2752-2754

[95] Kristensson E, Berrocal E, Richter M, Alden M. Nanosecond structured laser illumination planar imaging for single-shot imaging of dense sprays. Atomization and Sprays. 2010;**20**(4): 337-343

[96] Berrocal E, Kristensson E, Hottenbach P, Alden M, Grunefeld G. Quantitative imaging of a non-combusting diesel spray using structured laser illumination planar imaging. Applied Physics B: Lasers and Optics. 2012;**109**:683-694

[97] Desantes JM, Pastor JV, Pastor JM, Julia JE. Limitations on the use of the planar laser induced Exciplex fluorescence technique in diesel sprays. Fuel. 2005;**84**(18):2301-2315

[98] Melton LA, Verdieck JF. Vapor/ Liquid Visualization in Fuel Sprays. International Symposium on Combustion. 1985;**20**(1):1283-1290

[99] Murray AM, Melton LA. Fluorescence methods for determination of temperature in fuel sprays. Applied Optics. 1985;**24**(17):2783-2787

[100] Lemoine F, Antoine Y, Wolff M, Lebouche M. Simultaneous temperature and 2d velocity measurements in a turbulent heated jet using combined

laser-induced fluorescence and LDA. Experiments in Fluids. 1999;**26**:315-323

[101] Depredurand V, Miron P, Labergue A, Wolff M, Castanet G, Lemoine F. A temperature-sensitive tracer suitable for two-color laser-induced fluorescence thermometry applied to evaporating fuel droplets. Measurement Science and Technology. 2008;**19**(10):105103

[102] Castanet G, Lavieille P, Lebouché M, Lemoine F. Measurement of the temperature distribution within monodisperse combusting droplets in linear streams using two-color laser-induced fluorescence. Experiments in Fluids. 2003;**35**(6):563-571

[103] Omrane A, Sarner G, Alden M. 2d-temperature imaging of single droplets and sprays using thermographic phosphors. Applied Physics B: Lasers and Optics. 2004;**79**:431-434

[104] Yi SJ, Kim KC. Phosphorescence-based multiphysics visualization: A review. Journal of Visualization. 2014;**17**: 253-273

[105] Allison SW, Gillies GT. Remote thermometry with thermographic phosphors: Instrumentation and applications. The Review of Scientific Instruments. 1997;**68**(7):2615-2650

[106] Tran T, Kochar Y, Seitzman JM. Measurements of acetone fluorescence and phosphorescence at high pressures and temperatures. In: Paper AIAA-2006-831-356, 44th AIAA Aerospace Science Meeting. Reno, NV: AIAA; 2006

[107] Ritchie BD, Seitzman JM. Simultaneous imaging of vapor and liquid spray concentration using combined acetone fluorescence and phosphorescence. In: Paper AIAA-2004-0384, 42nd AIAA Aerospace Science Meeting. Reno, NV: AIAA; 2004

[108] Wigley G, Goodwin M, Pitcher G, Blondel D. Imaging and PDA analysis of a GDI spray in the near-nozzle region. Experiments in Fluids. 2004;**36**(4): 565-574

Section 3

Fundamental Research on Liquid Droplet Behaviors

Chapter 3

Behaviors of Multi-Droplets Impacting on a Flat Wall

Hongliang Luo and Feixiang Chang

Abstract

Microscopic characteristics of fuel spray are very important for atomization and mixture formation. The droplet size, number density, velocity distribution as well as minimum distance reveal the quality of spray and atomization, which affects the subsequent combustion and emissions for different engines such as vehicle, marine and aircraft. Moreover, in the internal combustion engine, the spray-wall impingement is difficult to avoid, which is the main source for soot emissions. Nowadays, regulations for emissions become straight by governments. Therefore, it is urgent for us to alleviate the energy and emissions crisis. In this study, the droplets behaviors will be characterized under the related engine working state. Firstly, the experimental setup and measurement were explained in detail. Then, images process method was induced to calculate the droplet size, velocity and distance among them. Finally, results of the impinging spray were presented. One thing should be noted, as the dense region is not available to detect the droplets by the observation. Therefore, a spray "slicer" was designed and applied to cut the spray slim. Finally, multi-droplets were generated, and the results can be concluded as well. All the results could provide insights into the impacting behaviors for better understanding the droplet dynamics.

Keywords: multi-droplets, impacting behaviors, energy conversion, microscale analysis

1. Introduction

Droplet-wall interactions are widely used in the gas-turbine and internal combustion engines under a range of different conditions [1]. After injected into the cylinder, droplets with various velocities travel along the spray axis to impact on the piston head or cylinder wall to form fuel film as well as breakup into small ones splashing off the wall [2]. Moreover, the rebounding and splashing droplets coalesce and collide with each other in the air by the help of vortex, which may impinge on the cylinder wall again or eventually adhere on it [3]. Especially, for the cold-start operation, the liquid film on the wall surface significantly contributes to the soot emissions [4]. Therefore, understanding the mechanism of impinging spray is essential for better design and optimization on the engine performance. Specially, the investigations on droplet characteristics could contribute to model the impinging behaviors for separating the deposit/splash droplets under a series of pressure and temperature conditions.

IntechOpen

Recently, droplets impacting on the wall becomes a hotpot for investigations, especially after Moreira et al. [5] put forward the question that "How much of single droplet impact research is useful?". As a fundamental study, the single drop-wall impacting can clearly explain the relationship between the splashing behavior and non-dimensional parameters, such as Reynolds, Weber, Ohnesorge and K numbers [6]. Previously, Walzel [7] firstly described the droplets impinging behavior. Then, the splashing characteristics was formulated by Yarin and Weiss [8]. Followed by them, Mundo et al. [9] firstly applied K number to separate droplets behaviors of "splashing" and "no-splashing". Then, Bai et al. [10] characterized the different impingement regimes of the incident drops by Weber number. Then it was refined and developed well by other scholars [11–13] Recently, Cen et al. [14] studied the dynamic break-up of the alternative biofuel impacting on the hot wall under the film boiling regime with different Weber number. They found that droplet breakup was dominated by Rayleigh instability, and the timing of jet break-up agreed well with the theory of Rayleigh instability. Qin et al. [15] investigated the splashing behaviors of an impacting drop on the hot surface to compare the effects of fuel viscosity and surface roughness. Liu et al. [16] studied the spreading dynamics of a single droplet impacting on a heated surface. They found that impacting behavior such as "depositing", "rebounding" and "breakup with atomization" were sensitive to both surface temperature and Weber number.

According to the literature surveys above, although many investigations were performed on the drop-wall interaction. It is still necessary to conduct an in-depth study on the multi-droplets wall-impingement. This study aims to fill the research gap. The target is to investigate the characteristics of multi-drops impingement on the wall experimentally under various conditions. Particle image analysis (PIA) technique was implemented to obtain the microscopic observations. Firstly, the spray droplets were checked by a single hole injector. Then, a spray "slicer" was introduced to make the multi-droplets impinging on the wall. Moreover, the thickness of "slicer" was defined. Finally, 0.4 and 0.04 mm in thickness of the "slicer" were determined to test at the spray tip and quasi-steady state.

2. Experimental system and image processing method

The experimental setup is shown in **Figure 1**. It consists of three parts, including an injection system, an optical system and one constant volume chamber. Fuel was injected into the chamber, then developed and impinged on a flat wall by a high-pressure common rail. The high-pressure constant volume chamber was equipped with four glass-windows to transmit the observation light. Similar to the shadow method, particle image analysis (PIA) was applied in this research, which was also used in our published literatures [17, 18]. Therefore, here only a brief introduction is shown. A double pulsed Nd: YAG laser of 532 nm wavelength was used to illuminate the spray. And microscopic behaviors of spray can be observed through a charge-coupled device (CCD) camera (Flowtech Research Inc., FtrNPC). Moreover, a microscope was used to connect the camera to obtain the tiny droplets clearly with three teleconverters (Kenko Tokina, N-AF 1.4X TELEPLUS MC4*3) connected to amplify the micro-image. Besides, owing to the dense drops along the spray axis, a house-designed spray "slicer" was designed and applied to get the slim spray, as depicted in **Figure 2**. The spray "slicer" made of steel with length of each slicer at 30 mm and height at 3 mm. And the distance between two slicer was defined as the thickness, which could be adjusted from 0.04 to 1 mm.

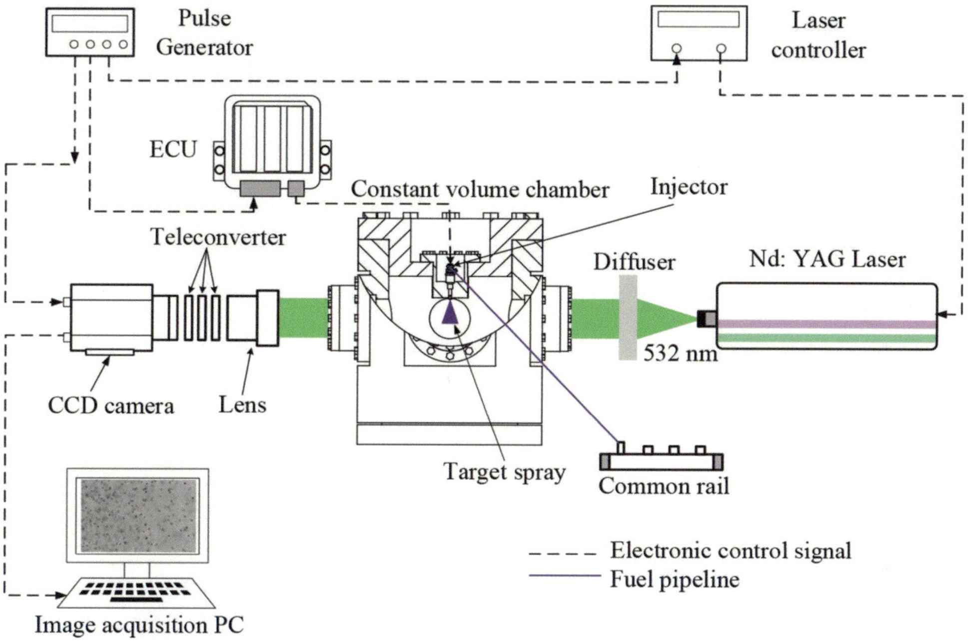

Figure 1.
Experimental apparatus.

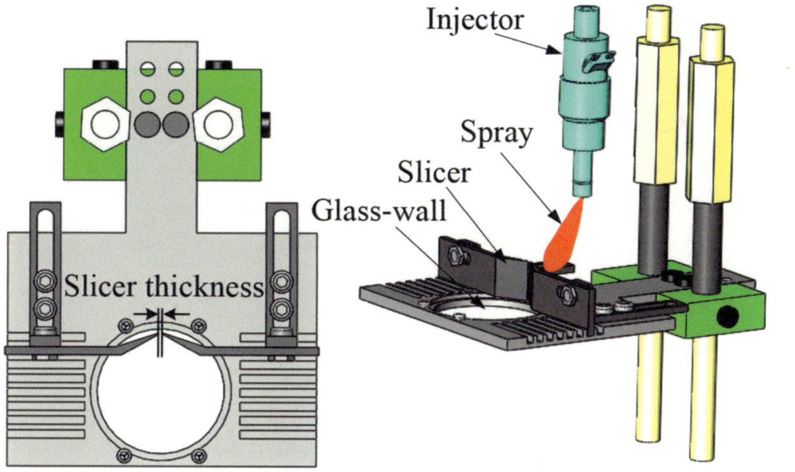

Figure 2.
Schematic of spray "slicer".

Table 1 lists the experimental conditions. On mini-sac injector with a single hole of 0.135 mm in diameter and 0.65 mm in hole length was used, resulting in the length-to-diameter (L/D) ratio at 4.8. Toluene was used as the tested fuel with density being 866 kg/m^3 under the room temperature. The properties such as viscosity and surface tension under the room temperature are 5.89 × 10–4 N·s/m^2 and 0.0285 N/m, respectively. The injection mass of toluene was kept constant at 4.0 mg. The chamber was filled with nitrogen to conduct the experiment under the non-evaporation condition. And the ambient pressure was changed between 0.1 and 0.5 MPa, leading to the gas

Injector Parameters	
Injector Type	**Mini-sac with a single hole**
Hole Length (mm)	0.65
L/D Ratio	4.8
Hole Diameter (mm)	0.135
Fuel Properties @ 298 K	
Test Fuel	Toluene
Density (kg/m^3)	866
Dynamic Viscosity ($N{\cdot}s/m^2$)	5.89×10^{-4}
Surface Tension (N/m)	0.0285
Conducted Conditions	
Injection Mass (mg)	4.0
Ambient Gas	Nitrogen
Injection Pressure (MPa)	10, 20, 30
Injection Duration (ms)	2.9, 2.1, 1.7
Ambient Temperature (K)	298
Ambient Pressure (MPa)	0.1 and 0.5
Ambient Density (kg/m^3)	1.95 and 5.95
Impinging Conditions	
Impingement Wall	Flat wall, quartz glass
Wall Roughness (μm)	Ra = 7.0
Wall Temperature (K)	298
Impingement Distance (mm)	22
Impingement Angel (deg)	45

Table 1.
Experimental conditions.

density at 1.19 and 5.95, respectively. Besides, the injection pressure was set among 10, 20 and 30 MPa with injection duration at 2.9,2.1 and 1.7 ms. A flat wall made of quartz glass was used as the impingement wall, as shown in **Figure 3**. The impingement distance from the nozzle exit to the wall was decided at 22 mm. And the impingement angle was determined at 45 deg. between the spray axis and the wall surface. One thing should be noted that the roughness of the flat wall was measured at Ra7.0 μm by a portable high-performance surface roughness and waviness measuring instrument (Kosaka Laboratory Ltd., SE300).

The image processing is shown in **Figure 3**. Firstly, the resolution can be calculated by the help of a micro-scale ruler. For filtering the spherical droplets, some refined criteria should be defined, such as roundness, pixel number and diameter. After detecting the spherical droplets, the distance between droplets can be obtained. Furthermore, with the interval timing between these two frames at 0.4 μs, the velocity of each droplet can be calculated. Moreover, in order to make the statistical results reasonable, at least 100 repeated tests were conducted under the same condition to make sure more than 5000 particles are available for statistics. Finally, the droplet

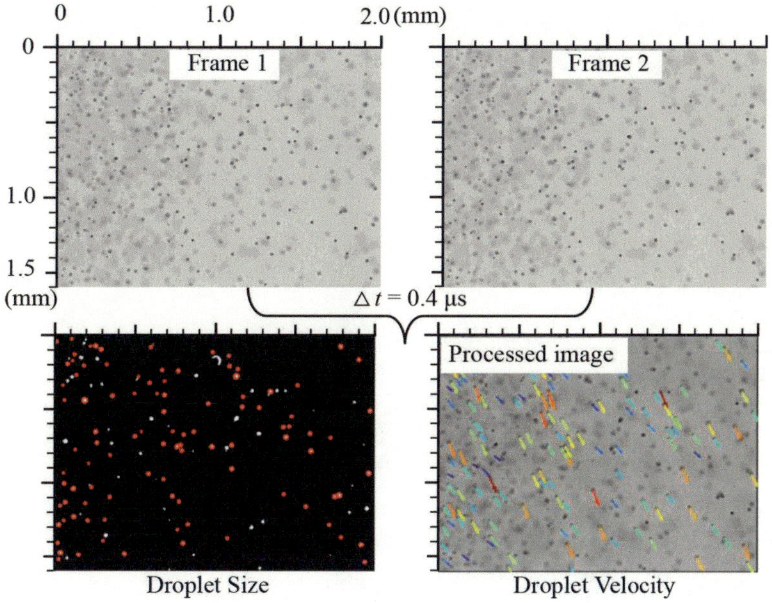

Figure 3.
Image processing.

size, velocity, distance and non-dimensional results can be gotten. The details about the algorithm can be seen in the article of Wang et al. [19].

3. Results and discussion

3.1 Impinging spray and droplet behaviors

Microscopic spray development before impingement under 0.1 and 0.5 MPa are presented in **Figure 4**. It can be seen that all the images are arranged along the spray axis and then rotated 45 deg. at counterclockwise connected. Under P_{amb} = 0.1 MPa, when fuel spray at the near field region, less droplets can be seen. With development at 2–4 mm from the origin, it is clear to see the liquid column. With longer penetration development at 8–10 mm, better atomization can be seen that liquid sheet located in the core with droplets at the periphery. At further locations, more ligaments and clear droplets can be captured. Besides, increased the ambient pressure to 0.5 MPa, more atomized droplets can be obtained because of the strong sheer force to promote droplets collision, leading to dense spray when compared to that of 0.1 MPa. In this case, although PIA could obtain the microscopic spray clearly at periphery, it is still difficult to clear observe the drops at the dense spray regions.

Then, typical morphologies at (10, 10) and (14, 14) are shown in **Figure 5**. Before impingement at (10, 10), both ligaments and droplets can be seen at 0.1 MPa due to the primary breakup. Under 0.5 MPa, intense interactions result in more small size droplets. Transferred to being impingement location at (14, 14). The white solid line represents the surface of the flat wall. At the near wall region, it is too dense to identify the droplets along the spray axis when compared to (10, 10), resulting in only droplets at periphery can be observed.

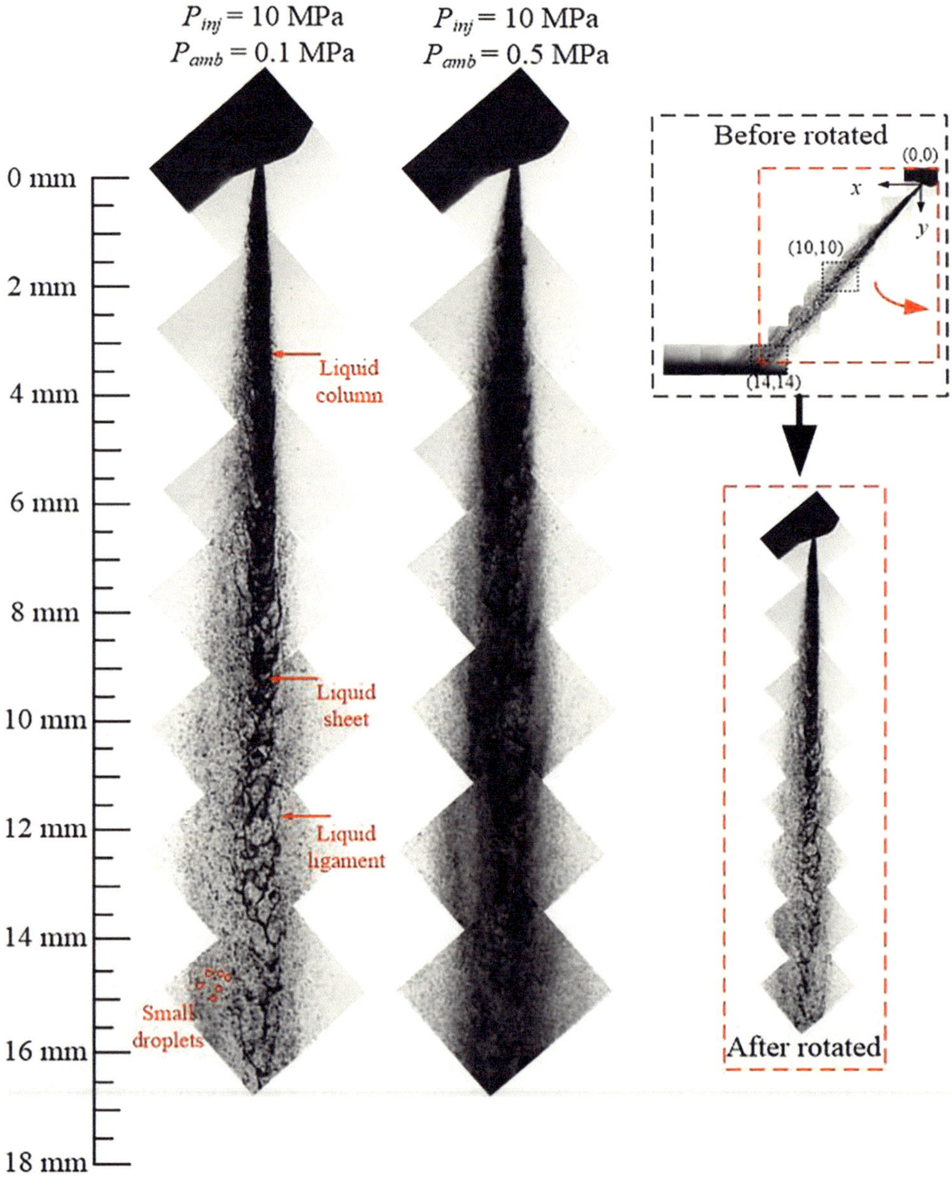

Figure 4.
Microscopic behaviors of spray and droplets.

Generally, the atomized droplets are the outcomes of competitions between aerodynamic force and surface tension. The friction in the gas-fuel interface results in the disintegration and instable deformation, finally contributing to the formation of smaller droplets. Therefore, We_g can be used to describe instability of the droplet formations and predicts the behaviors of droplet breakup furtherly, defined as:

$$We_g = \frac{\rho_g v^2 d}{\sigma} \quad (1)$$

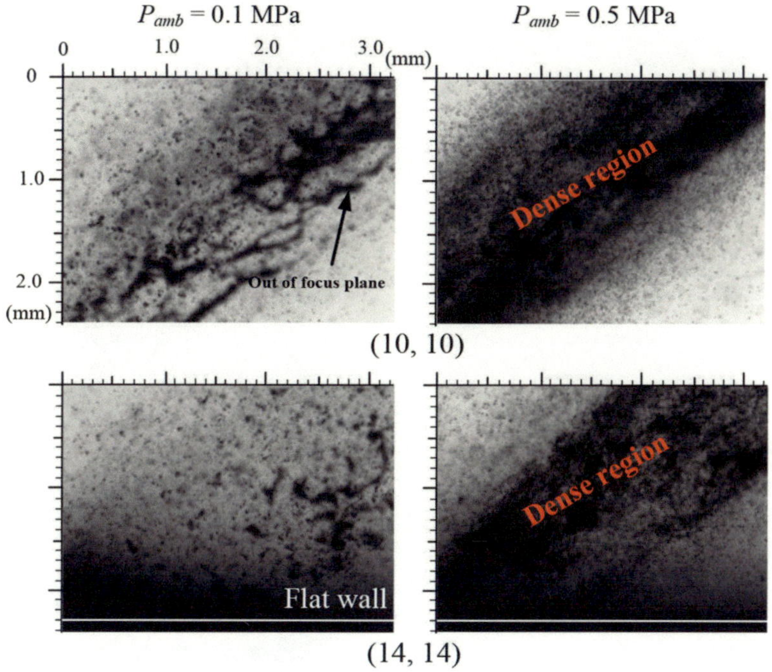

Figure 5.
Spray morphology under different ambient pressures.

where ρ_g is the ambient gas density with σ being the liquid surface tension. v is the relative velocity between droplet and ambient gas. d represents the droplet diameter.

Previously, Wierzba [7] found the vibrational breakup occurs at a relatively low We_g, and droplets are easy to break up into smaller ones. Therefore, We_g can be applied to indicate the degree of atomization during development, which is divided into four parts: when $0<We_g<1$ and $1<We_g<10$, it suggests good atomization; when $10<We_g<14$, the transient section indicates the further atomization can be expected; when $We_g>14$, the large droplets still exist.

The diameter-velocity distribution is presented in **Figure 6**. The scatter number in each sub-figure suggests the number of available droplets. The horizontal axis is droplet size (6–80 μm), and vertical axis is velocity (0–200 m/s). At 0.1 MPa, compared between (10, 10) and (14, 14) in Figures. (a) and (b), results are similar. However, more droplets with larger We_g can be seen when ambient pressure increased to 0.5 MPa in Figures. (c) and (d), suggesting these droplets have the potential for further atomization. Besides, in contrast to Figures. (c) and (d), the number of large We_g decreases significantly, indicating that higher ambient pressure favors fuel breakup and atomization. In other words, it can be concluded that fuel spray development is the process of We_g decreasing with better atomization.

The number probability of droplet size is calculated in **Figure 7**. Before impingement (10, 10), two curves are similar but with different peak values, and the peak value of 0.1 MPa is larger than that of 0.5 MPa. However, difference becomes not obvious. Furthermore, the droplet size corresponding to the peak value of curves is the same at 12 μm. One possible reason may be that the scale of atomization is mainly decided by the injector itself, resulting in less relationship with the ambient pressure. In order to discuss the velocity distribution, the probability of velocity is presented in

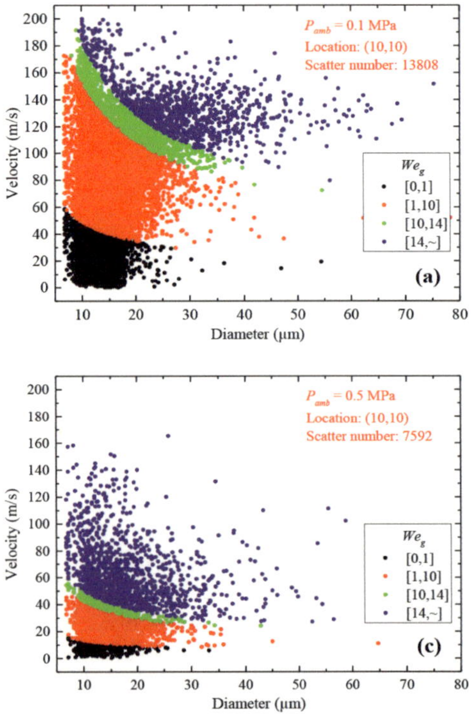

Figure 6.
Diameter–velocity distribution with gas phase weber number.

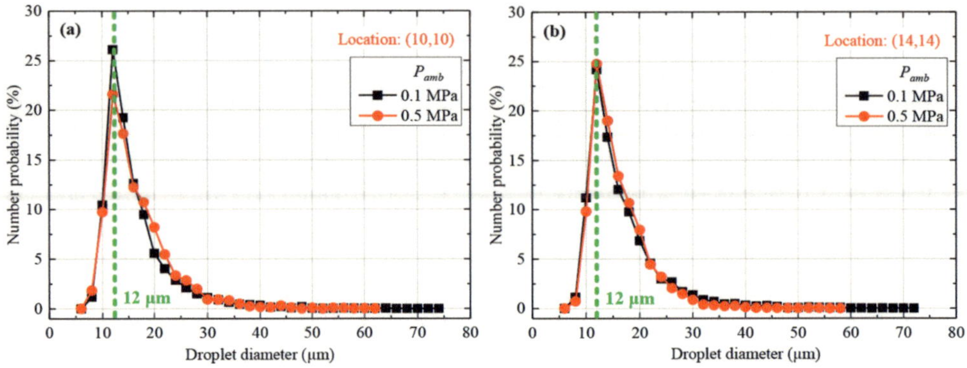

Figure 7.
Droplet diameter probability curves.

Figure 8. With the increase in ambient pressure, the maximum velocity reduces from 200 to 150 m/s. Besides, the velocity corresponding to the peak value decreases at elevated ambient pressure decreases from 65 to 25 m/s at (10, 10), from 45 to 10 m/s at (14, 14). Besides, different to droplet size distribution, the peak value in velocity decreases sharply under high ambient pressure owing to more energy dispersion.

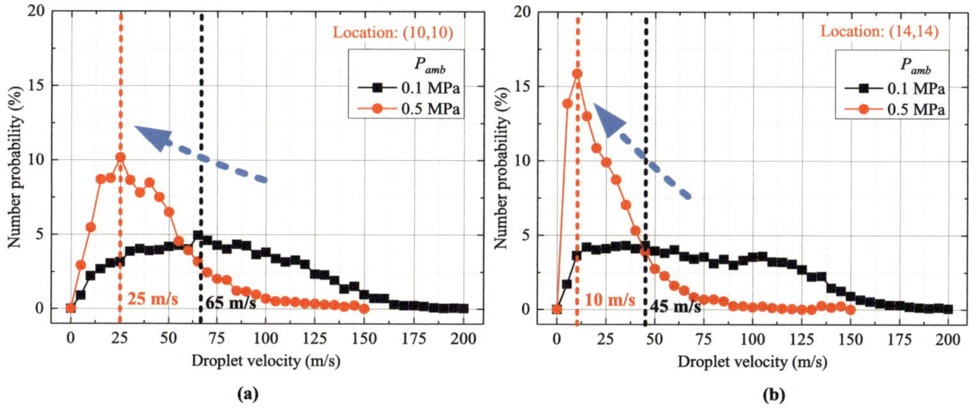

Figure 8.
Droplet velocity probability.

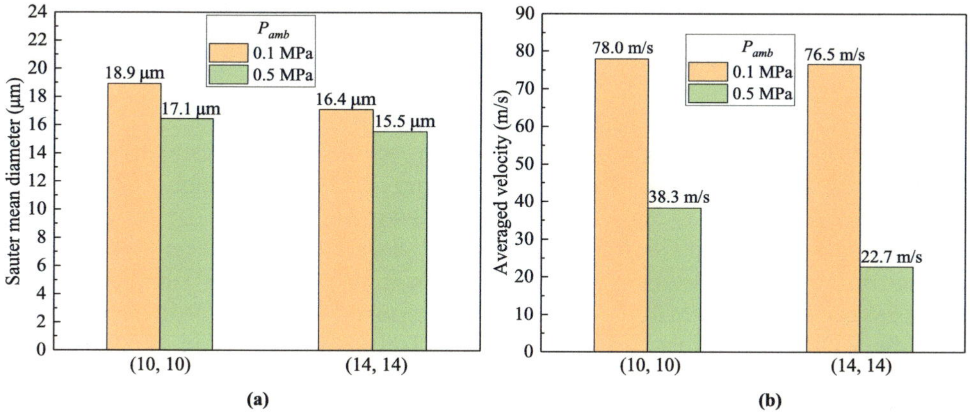

Figure 9.
(a) SMD and (b) averaged velocity.

All in all, it can be seen that velocity is sensitive to the location and ambient pressure significantly compared to diameter. And both ambient pressure and spray development decelerate droplets.

Sauter mean diameter (SMD = ∑D3/D2) is defined as the diameter of a droplet with the same ratio of volume to surface area, which is widely applied to evaluate the atomization, as shown in **Figure 9**. Here, SMD and averaged velocity under each condition are compared. SMD decreases with elevated ambient pressure and spray development, resulting in better atomization. The ambient shear force and high gas density should be the main reasons for it. Moreover, the averaged velocity decreases as well because of the resistance force. And the minimum distance of the selected droplets is defined as the shortest distance among them, as shown in **Figure 10**, which can be applied to represent the spray dispersion. The results of mean minimum distance (MMd) are calculated and depicted here. Similar to SMD and averaged velocity, MMd decreases with ambient pressure increasing but with different reason. That high ambient pressure compacts droplets, making them together due to the high ambient density should be one possible reason for it.

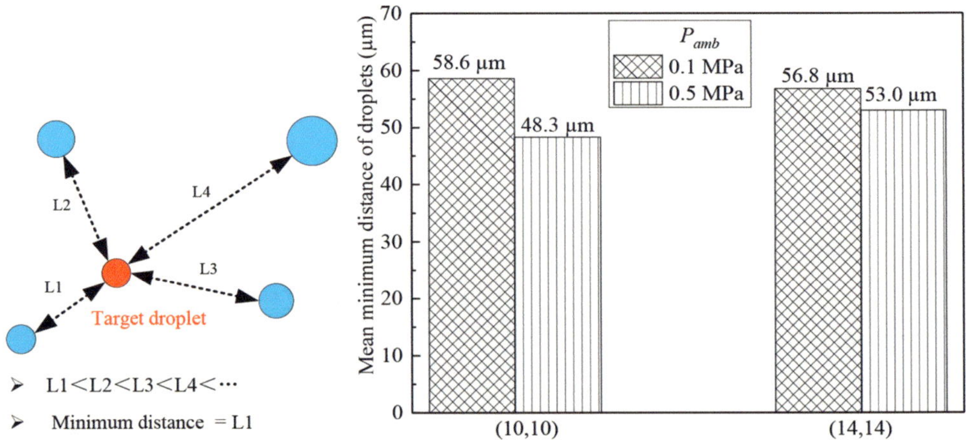

Figure 10.
Definition and results of mean minimum distances.

3.2 Sliced spray behaviors

In this part, the spray behaviors will be discussed at the spray tip for clear observation. Moreover, the spray "slicer" was applied here with the thickness at 0.4 mm.

Firstly, **Figure 11** shows the morphology of spray tip at two locations under the injection pressure of 10 and 30 MPa, respectively. At (10, 10), clear droplets, ligaments and even liquid sheets can be seen, indicating the spray slicer improves the atomization. Moved to the impinging location at (14, 14), the splashing crown structures can be seen clearly, which previously reported in the single drop impacting experiment [20, 21]. Besides, the photographing timing is recorded in each image. With an increase in injection pressure, shorter timing is needed due to high momentum energy from the drops.

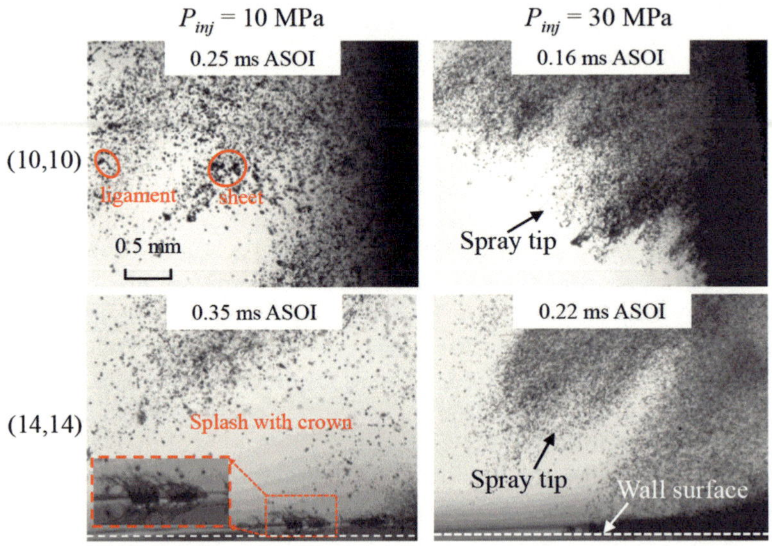

Figure 11.
Spray morphology under different injection pressures.

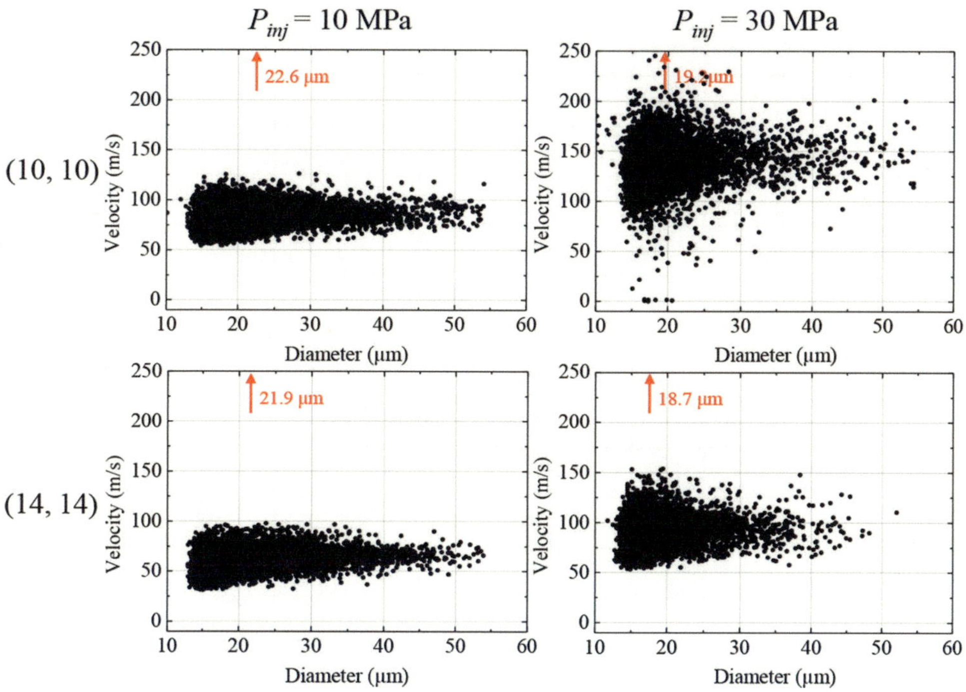

Figure 12.
"Fish-shape" distribution.

Droplet diameter-velocity distribution is presented in **Figure 12**. The horizontal axis is diameter, and vertical axis is velocity. Besides, SMD is depicted in each image. It is interesting to find the "fish-shape" distribution under all conditions with big "head" and small "tail". Furthermore, increased the injection pressure from 10 to 30 MPa, distribution becomes wider with higher velocity because of high kinetic energy. In contrast to location effect, the influence of injection pressure plays a much more important role on not only velocity but droplet size. It is evident that increasing injection pressure reduces SMD significantly, when compared it between (10, 10) and (14, 14).

The size probability is presented in **Figure 13**. It is interesting to find the almost half droplet diameter is in the range of 12–20 μm. Moreover, the peak value corresponding to diameter locates in only 12–16, indicating the droplet size is determined largely by the injector itself, instead of the injection pressure. Besides, the peak value increases with spray tip movement and injection pressure owing to better atomization. Furthermore, injection pressure shows lager effect on droplets atomization than that of measurement location.

Figure 14 shows the velocity distribution along x and y directions to describe both velocity and direction. Four quadrants are separated to define the velocity direction. Quadrant I indicates the incident droplet moving along the spray axis; Quadrant II indicates droplet splashing off the wall in the direction perpendicular to spray axis; Quadrant III indicates droplet rebounding into the air along the spray axis; Quadrant IV indicates droplet rushing to the wall in the completely opposite direction of Quadrant II. Moreover, the absolute velocity is the distance from one certain dot to the origin. Because of the spray tip, all the droplets show similar rushing direction to the wall in Quadrant I. And almost the same shape can be seen under 10 MPa. But increasing the injection pressure to 30 MPa, droplet velocity can reach to 200 m/s.

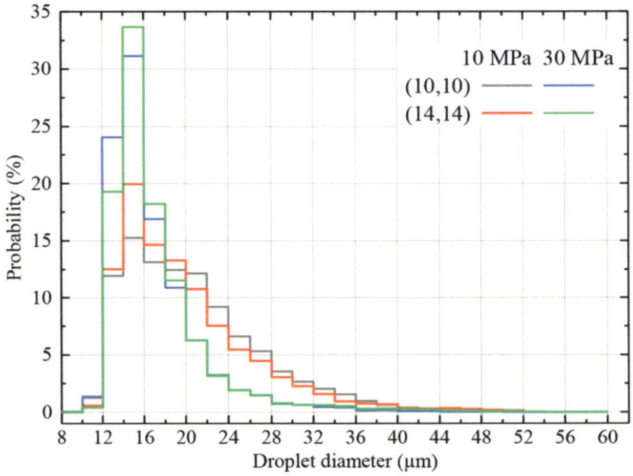

Figure 13.
Droplet size distribution.

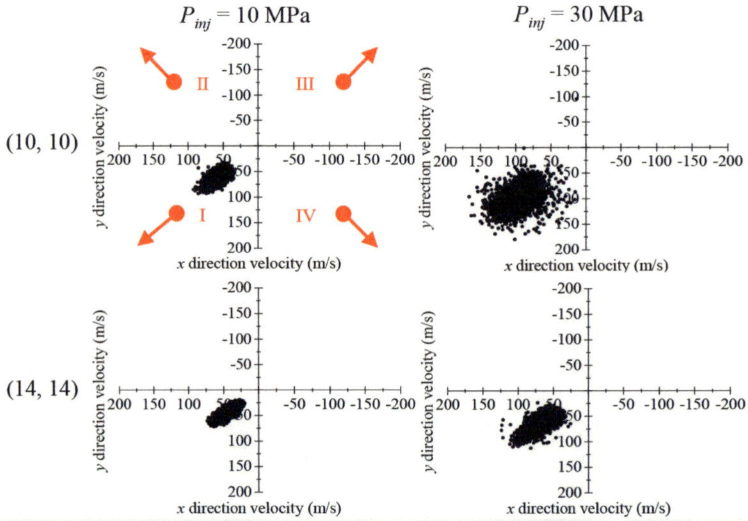

Figure 14.
Droplet velocity direction.

Figure 15 shows the diameter-minimum distance. As explained before, minimum distance (Md) could be used to judge the droplet dispersion with larger Md to indicate the better dispersion. All results show similar distribution with little difference. One thing can be concluded from distributions that increased the injection enlarges Md at a certain level. Therefore, in order to discuss the Md clearly, mean minimum distance (MMd) is used in **Figure 16** as well as averaged velocity. For averaged velocity, increasing injection pressure accelerates droplet greatly from 83.3 to 137.0 m/s at (10, 10). But this increasing tendency becomes moderate at (14, 14). In addition, averaged velocity decreases slightly with tip movement due to the air drag force. For MMd, both movement and injection pressure enlarge MMd, suggesting better atomization. However, one interesting thing is that different to velocity, MMd is more sensitive to the location instead of injection pressure. Droplets disperse well with the spray development should be one possible reason for it.

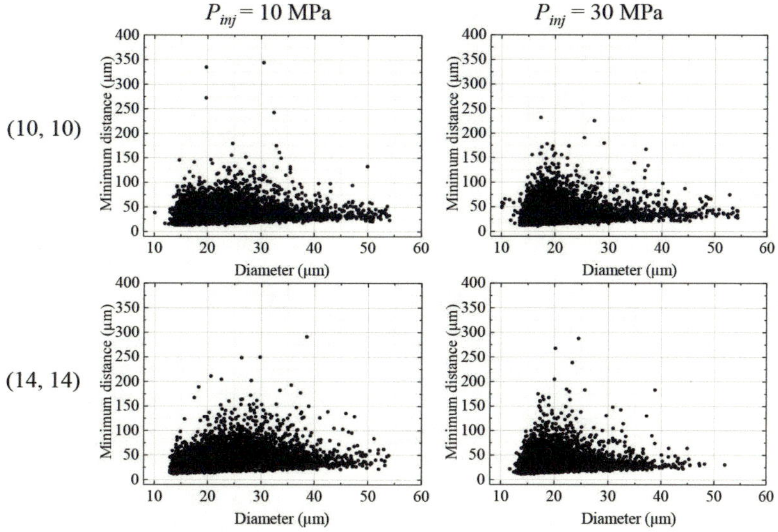

Figure 15.
Droplet diameter-Md distribution.

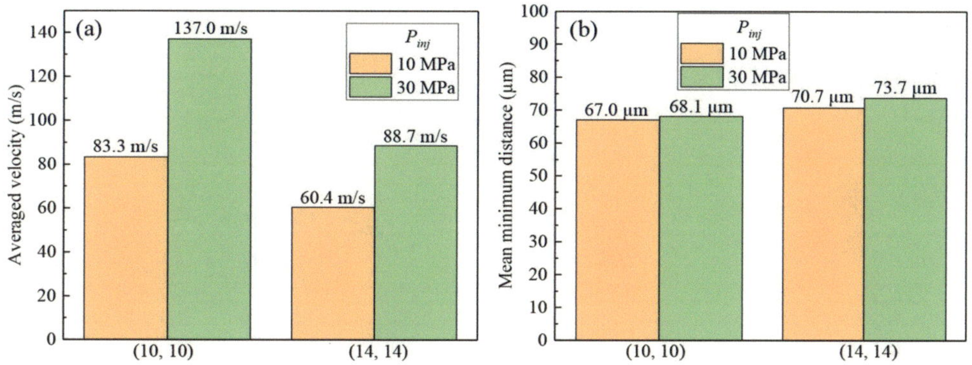

Figure 16.
(a) Averaged velocity and (b) MMd.

3.3 Multi-droplets impacting characteristics

In order to explore the multi-droplets impacting behaviors deeply, the thickness of "slicer" was decreased to 0.04 mm, as shown in **Figure 17**. At spray tip, droplets can be recognized at thickness of 0.4 and 0.1 mm. However, changed to quasi-steady state, the thickness should be decreased much more. Finally, the thickness was determined at 0.04 mm. And in this part, multi-droplets can be achieved through the "slicer". Moreover, the impacting and splashing behaviors of droplets will be discussed in detail.

Figure 18 presents the diameter-velocity distribution at (16, 14) and (18, 14). The horizontal axis is the droplet diameter, with vertical axis being as the velocity. Results of 10, 20 and 30 MPa are depicted by the black, red and blue data, respectively. For the velocity, at (16, 14), the velocity ranges from 0 to 110 m/s. But it decreases to only 50 m/s when the location is transferred to (18, 14). And it is supposed that the energy

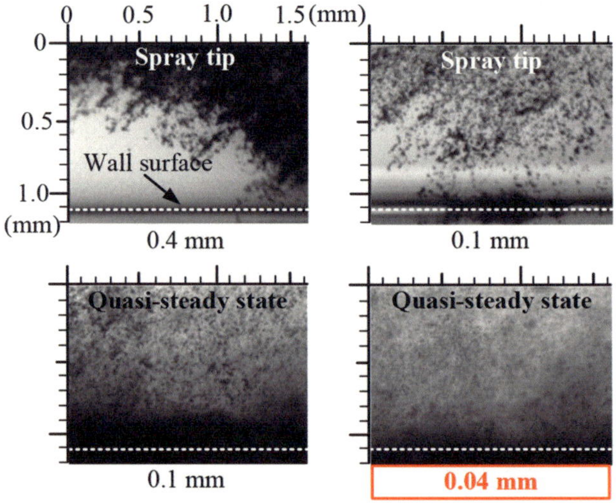

Figure 17.
Comparison in the clear observations with different thickness of "slicer".

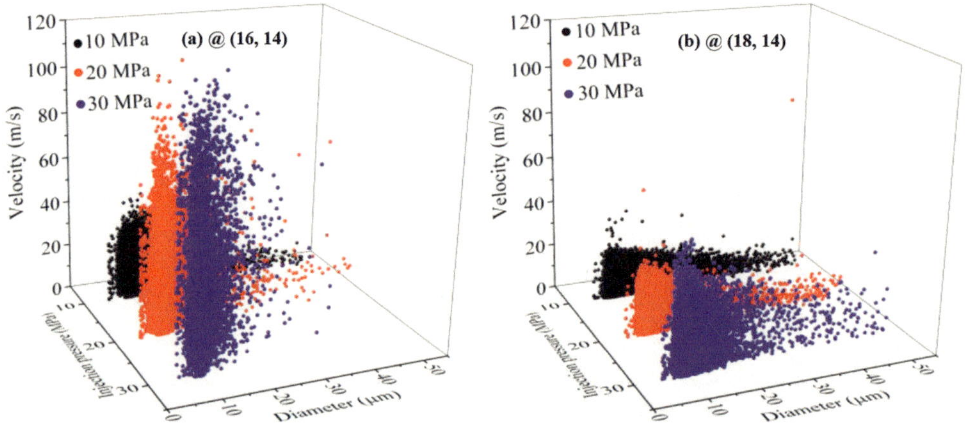

Figure 18.
Droplet diameter-velocity distribution.

disperses with the interaction of droplets and fuel film on the wall during the injection, which will be discussed more in the following part. Moreover, for the droplet size, it is clear to see that diameter increases significantly when transferring from (16, 14) to (18, 14). Droplets collision with each other after impacting on the wall should be one possible reason for it.

In order to analyze the size distribution statistically, the probability of the diameter is applied and shown in **Figure 19**. At (16, 14), the peak value corresponding to the diameter increases with higher injection pressure from 20–25%. In contrast to (18, 14), peak value declines obviously, suggesting that droplet size increases from (16, 14) to (18, 14) with spray movement but decreases with increasing injection pressure. Finally, SMD is calculated and depicted in **Figure 20**. It shows that SMD decreases slightly with an increase in injection pressure at (16, 14). However, larger influence of injection pressure can be seen at (18, 14). The larger-size drops are easy to break up into smaller ones owing to high kinetic energy. While, by comparing the

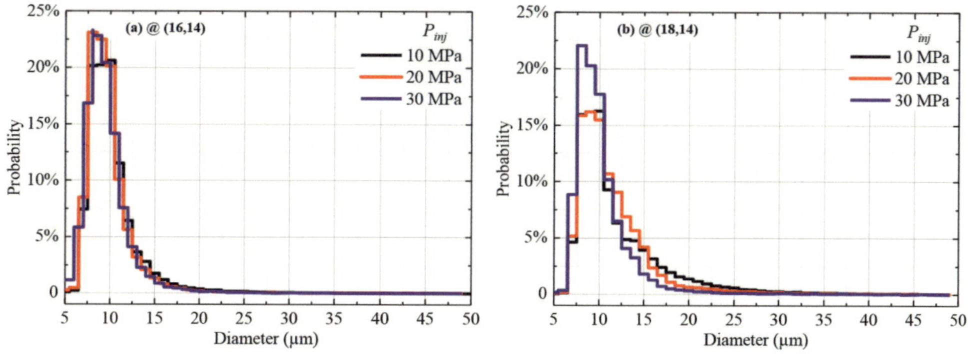

Figure 19.
Probability of diameter at different locations. (a) @(16, 14) (b) @(18, 14).

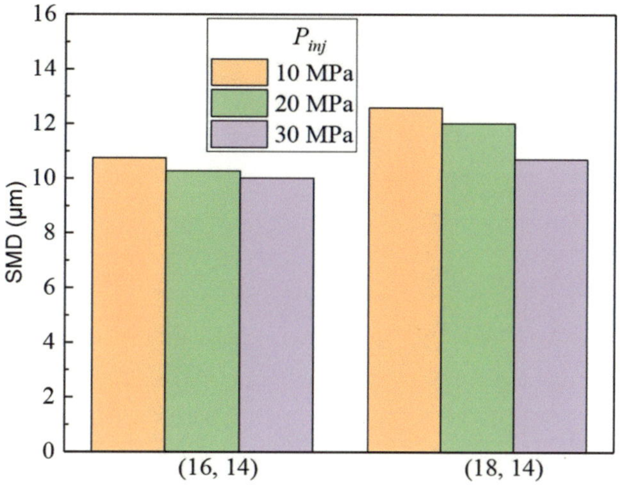

Figure 20.
SMD at different locations.

location effect, SMD is enlarged with movement. After impingement on the wall, the splashing droplets may collide and coalesce by others can explain this phenomenon. More details will be discussed through the velocity analysis.

Figure 21 shows the velocity-direction distribution. X and Y coincide with the direction presented in **Figure 4**. And the distance from the origin to one selected droplet presents the magnitude velocity. At (16, 14), almost all the droplets rush into the wall along the spray jet direction. Besides, high injection pressure accelerates droplets. While, changed to (18, 14), droplets change direction in Quadrant II, indicating the splashing direction. And this is the main reason for the low velocity and large size. After impacting on the wall, droplets splashing off the wall results in energy loss. Additionally, some droplets may collide and change the direction as well as coalescence into larger ones. One thing should be pointed that at (16, 14) and (18, 14), droplets locate in all these four quadrants because of the vortex.

Next, the averaged velocity was calculated and presented in **Figure 22**. It shows that at (16, 14), velocity increases significantly with higher injection pressure. Compared with after impingement at (18, 14), only half velocity can be seen due to

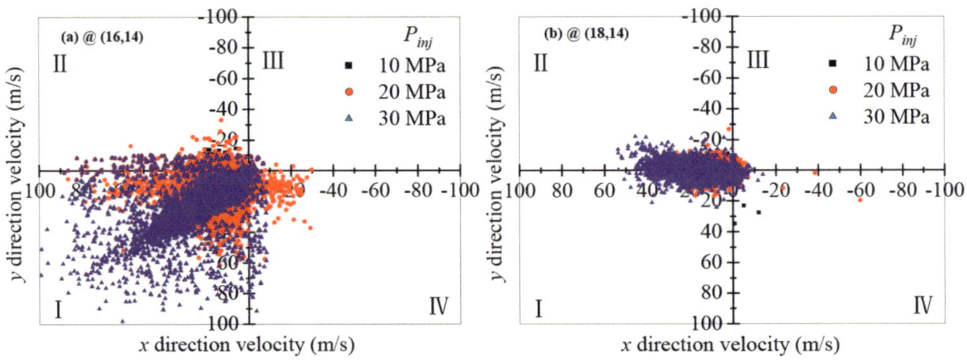

Figure 21.
Droplet velocity distribution at different locations.

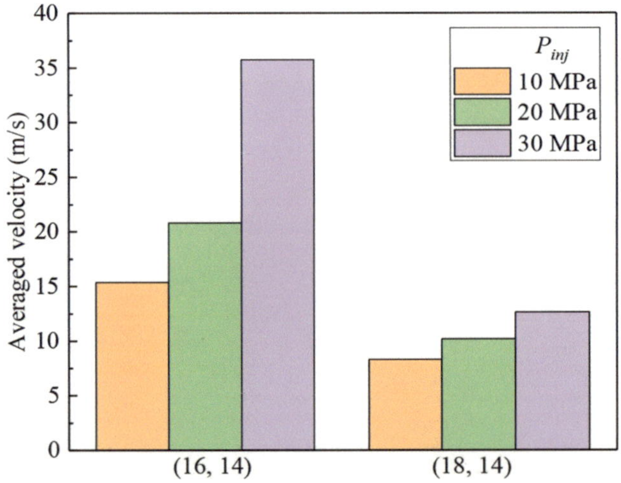

Figure 22.
Averaged velocity at different locations.

the interaction consuming energy. Finally, the mechanisms of multi-droplets impacting on the wall is conceived. After impingement, the secondary droplets are generated by the irregular crown structure with the help of the interaction between fuel film and droplets. On one hand, the splashing droplets after impingement collide with each other and then coalesce into larger one. On the other hand, the splashing crown structures also collide owing to the multi-droplets crash into the fuel film, leading to droplets generated from the rim of the crown body. Moreover, the crown bodies collide, then generate droplets splashing off the structure to increase the totally calculated SMD.

4. Conclusions

In this study, droplet behaviors were investigated before, during and after impingement on the wall. For clear observation, free and impinging spray (with/without "slicer") at different thickness of 0.4 and 0.04 mm were characterized to evaluate the droplet size, velocity and distance. Besides, different influences including measurement locations, injection pressures and ambient pressures were considered to better understand the behaviors of multi-droplets impacting on the wall. The following conclusions can be drawn as:

1. We_g can be applied to evaluate the degree of breakup and atomization. Owing to the droplets breakup and atomization, We_g decreases from (10, 10) to (14, 14). Moreover, high ambient pressure can promote this phenomenon. Furthermore, ambient pressures and locations has less effect on the scale of atomization as the diameter corresponding to peak value of probability is constant. However, velocity is more sensitive to the ambient pressure and locations compared with the droplet size.

2. Higher injection pressure widens the distribute in velocity but compacts the diameter distribution. Droplet velocity direction can be obtained and mapped in four quadrants. And Md of droplets is defined and employed to evaluate the droplets dispersion. Higher injection pressure decreases SMD, accelerates droplets and promotes the droplets dispersion owing to high kinetic energy and better atomization.

3. After impacting on the wall, SMD increases but velocity decreases. The multi-droplets impacting on the wall cannot be considered as the superposition of undisturbed single droplet. Not only splashing droplets but also splashing crown structures collide and coalesce to generate secondary droplets with large size and low velocity.

Acknowledgements

The author would like to acknowledge National Natural Science Foundation of China [Grant 51909037] and Foundation of State Key Laboratory of Engines [No. K2022-12].

Conflict of interest

The authors declare no conflict of interest.

Author details

Hongliang Luo[1,2]* and Feixiang Chang[2]

1 Foshan University, Foshan, P.R. China

2 Hiroshima University, Higashi-Hiroshima, Japan

*Address all correspondence to: luo@hiroshima-u.ac.jp

References

[1] Luo H, Uchitomi S, Nishida K, Ogata Y, Zhang W, Fujikawa T. Experimental investigation on fuel film formation by spray impingement on flat walls with different surface roughness. Atomization and Sprays. 2017;**27**:611-628. DOI: 10.1615/AtomizSpr.2017019706

[2] Hiroyasu H, Arai M. Structures of fuel sprays in diesel engines. SAE Technical Paper. 1990:900475. DOI: 10.4271/900475

[3] Luo H, Nishida K, Uchitomi S, Ogata Y, Zhang W, Fujikawa T. Microscopic behavior of spray droplets under flat-wall impinging condition. Fuel. 2018;**219**:467-476. DOI: 10.1016/j.fuel.2018.01.059

[4] Luo H, Nishida K, Uchitomi S, Ogata Y, Zhang W, Fujikawa T. Effect of temperature on fuel adhesion under spray-wall impingement condition. Fuel. 2018;**234**:56-65. DOI: 10.1016/j.fuel.2018.07.021

[5] Moreira ALN, Moita AS, Panao MR. Advances and challenges in explaining fuel spray impingement: How much of single droplet impact research is useful? Progress in Energy and Combustion Science. 2010;**36**:554-580. DOI: 10.1016/j.pecs.2010.01.002

[6] Cossali GE, Coghe ALDO, Marengo M. The impact of a single drop on a wetted solid surface. Experiments in Fluids. 1997;**22**:463-472. DOI: 10.1007/s003480050073

[7] Wierzba A. Deformation and breakup of liquid drops in a gas stream at nearly critical weber numbers. Experiments in Fluids. 1990;**9**:59-64. DOI: 10.1007/BF00575336

[8] Yarin AL, Weiss DA. Impact of drops on solid surfaces: Self-similar capillary waves and splashing as a new type of kinematic discontinuity. Journal of Fluid Mechanics. 1995;**283**:141-173. DOI: 10.1017/S0022112095002266

[9] Mundo CHR, Sommerfeld M, Tropea C. Droplet-wall collisions: Experimental studies of the deformation and breakup process. International Journal of Multiphase Flow. 1995;**21**:151-173. DOI: 10.1016/0301-9322(94)00069-V

[10] Bai CX, Rusche H, Gosman AD. Modeling of gasoline spray impingement. Atomization and Sprays. 2002;**12**:1-27. DOI: 10.1615/AtomizSpr.v12.i123.10

[11] Sivakumar D, Tropea C. Splashing impact of a spray onto a liquid film. Physics of Fluids. 2002;**14**:85-88. DOI: 10.1063/1.1521418

[12] Kalantari D, Tropea C. Spray impact onto flat and rigid walls: Empirical characterization and modelling. International Journal of Multiphase Flow. 2007;**33**:525-544. DOI: 10.1016/j.ijmultiphaseflow.2006.09.008

[13] Roisman IV, Breitenbach J, Tropea C. Thermal atomisation of a liquid drop after impact onto a hot substrate. Journal of Fluid Mechanics. 2018;**842**:87-101. DOI: 10.1017/jfm.2018.123

[14] Cen C, Wu H, Lee CF, Liu F, Li Y. Experimental investigation on the characteristic of jet break-up for butanol droplet impacting onto a heated surface in the film boiling regime. International Journal of Heat and Mass Transfer. 2018;**123**:129-136. DOI: 10.1016/j.ijheatmasstransfer.2018.02.059

[15] Qin M, Guo Y, Tang C, Zhang P, Huang Z. Spreading and bouncing of liquid alkane droplets upon impacting on a heated surface. International Journal of Heat and Mass Transfer. 2020;**159**:120076. DOI: 10.1016/j.ijheatmasstransfer.2020.120076

[16] Liu H, Si C, Cai C, Zhao C, Yin H. Experimental investigation on impact and spreading dynamics of a single ethanol–water droplet on a heated surface. Chemical Engineering Science. 2021;**229**:116106. DOI: 10.1016/j.ces.2020.116106

[17] Luo H, Jin Y, Nishida K, Ogata Y, Yao J, Chen R. Microscopic characteristics of impinging spray sliced by a cone structure under increased injection pressures. Fuel. 2021;**284**:119033. DOI: 10.1016/j.fuel.2020.119033

[18] Luo H, Nishida K, Ogata Y, Zhang W. Microscopic characteristics of near-nozzle spray at the initial and end stages. Fuel. 2021;**283**:118953. DOI: 10.1016/j.fuel.2020.118953

[19] Wang C, Wang Y, Zhu J, Long W, Nishida K. Design and application of an algorithm for measuring particle velocity based on multiple characteristics of a particle. Fuel. 2020;**271**:117605. DOI: 10.1016/j.fuel.2020.117605

[20] Krechetnikov R, Homsy GM. Crown-forming instability phenomena in the drop splash problem. Journal of Colloid and Interface Science. 2009;**331**(2):555-559. DOI: 10.1016/j.jcis.2008.11.079

[21] Rioboo R, Bauthier C, Conti J, Voue M, DeConinck J. Experimental investigation of splash and crown formation during single drop impact on wetted surfaces. Experiments in Fluids. 2003;**35**(6):648-652. DOI: 10.1007/s00348-003-0719-5

Chapter 4

Interaction and Transport of Liquid Droplets in Atmospheric Pressure Plasmas (APPs)

Muhammad M. Iqbal and Mark M. Turner

Abstract

The transport of liquid droplets in atmospheric pressure plasma (APP) has been recognized as a two-phase flow. The liquid droplet introduces a perturbation in APP and under several constraints, the behavior of this perturbation is not clear during transport. As soon as the droplets interact with the APP, they become charged, which causes the droplets to evaporate. Using 1D normalized fluid model, we first describe how the liquid droplets charge and interact with helium APP. The impact of this droplet-plasma interaction is then discussed and described using 2D coupled fluid-droplet model. The numerical modeling outcomes suggested that the evaporation of droplets has emerged as a primary mechanism in plasma; however, the mutual interactions, such as grazing and coalescence between the droplets, are dominant at higher precursor flow rates (>100 μl min^{-1}). To demonstrate the importance of different liquid precursors during droplet-plasma interaction, we analyzed the spatiotemporal patterns of discharge plasma while considering the effects of HMDSO, n-hexane, TEOS, and water. Variable gas and liquid precursor flow rates are used to further examine the features of discharge plasma. Comparing penning ionization to other ionization processes, it is found to be the prime activity along the pulse of droplets, demonstrating the significance of small nitrogen impurities. Using a laser diffraction particle size analysis approach as part of an APP jet deposition system, the validation of the numerical simulations is proven by comparison with experimental observations of droplet size distributions.

Keywords: atmospheric pressure plasma, droplet charging, evaporation, interaction, droplet model, two-phase flow, precursor flow, coupled model, grazing, coalescence, helium gas, N_2 impurities, penning ionization, deposition system, plasma simulations

1. Introduction

It has been widely accepted that atmospheric pressure discharge plasmas are recognized as an ideal source for achieving advanced desired requirements of coatings in industry, including surface modification, plasma-polymerized thin films, and engineering of mechanical components. Using nonequilibrium atmospheric pressure plasmas (APPs) with spraying precursors, recent studies in biology and medical science provided a clear path toward further innovation and improvement in these fields [1–5]. High-pressure discharges have been shown to be extremely useful for industrial applications such as sterilization of medical instruments, biomaterials for surface modifications, tissue engineering, blood coagulation, and therapy of skin disorders. While these applications operate at atmospheric pressure, the impact of small traces of impurities, such as gases, liquid vapors, and dust particles, cannot be ignored during processing because the interplay between these elements and their ramifications is controlled by the characteristics of the carrier gas mixtures [6–12]. Through the combination of liquid precursors in gas mixtures and discharge plasma, it is possible to produce material surfaces with desired properties. A direct injection of liquid precursors into discharge plasma can further complicate the occurring of chemical processes due to the shift from uniphase to two-phase flow. During this activity, two types of monomers were introduced into discharge plasma, such as vapors and liquid aerosols. Using both of these as a catalyst, the surface furnishings of the substrate were altered at a nanometer scale [13, 14].

Experiments perform an important role to test the theoretical concepts and, in some cases, also help to improve a perception with the use of outcomes from both fields. Using numerical models, this chapter addresses distinct characteristics of two-phase flow, including mixing, ignition, chemical reactions, and heat transfer in the chamber by exploring various aspects of two-phase flow [15, 16]. By investigating the spray dryers containing flue gases from coal fire power plants and diesel engine fuel, O′ Rourke and Dukowicz [17, 18] developed a numerical model of a multiphase flow which demonstrates the interaction of gas with liquid droplets. When a droplet comes in contact with the surface, its functionality can be altered in several ways, such as by altering the size, radius, morphology, and temperature of the droplet. Many studies on two-phase flow have been performed that examined both gas mixtures and droplets in a closely coupled plasma-droplet environment involving such factors as geometry, gas and precursor flow rates, temperature, and injection of mono- and polydisperse droplet sizes [19–22]. Droplets play a powerful role in a long discharge plasma channel due to their strong combination of external and internal forces, although sheath formation, around the droplet, has profound effects on how carriers are distributed in the plasma [23, 24].

The direct introduction of liquid precursors into atmospheric pressure plasma jets has proved an effective method for many applications in the past [2, 9, 25, 26], from industrial plasma processes to medical treatments, for instance, plasma-enhanced chemical vapor deposition, sterilization of medical equipment, and direct treatment of skin diseases. In [27], Navier–Stokes and energy equations have been solved to explain droplet deformation, solidification, and energy transfer on the substrate. It might be desirable for smooth plasma surface deposition to evaporate droplets completely in the plasma during transport, and particularly undesirable for partially evaporated droplets to reach substrate surfaces [28]. Surface deposition has numerous physical and chemical properties influenced by a variety of characteristics, especially the precursor and gas flow rates and the composition of the gas mixture, that is primarily responsible for the performance of the coating because a variety of chemical interactions take place during evaporation of droplets as well as particulate nucleation [19]. It has been determined that penning ionization occurs in the presence of trace amounts of nitrogen impurities in helium gas mixtures [29]. Previous research showed that Radio-frequency (RF) plasma sources ranging from 0.1 to 100 kHz produce satisfactory and promising results in controlling the relevant features of APP during processing within the last two decades [2, 30–32].

The bulk of the literature deals with gas dynamics in the presence of liquid droplets and the authors in [17, 18] elucidated the behavior of evaporation of droplets as well as mutual interactions between them in the case of small Weber numbers during liquid–gas interaction. In experiments, PlasmaStream™ atmospheric pressure plasma jet deposition systems [14] can be adapted to achieve uniform deposition of coatings on silicon substrate surfaces by adjusting the flow rate of liquid precursors. Inductively coupled plasmas are revitalized by the influence exerted by droplet breakup, desolvation, and coalescence, as demonstrated in [21]. In order to explore the vital characteristics of interaction among liquid droplets and flowing gas in the peculiar applications of combustion, the Eulerian–Lagrangian method is applied, and these features have been verified by the spectroscopic observations involving signal fluctuations in [15, 33–35]. Researchers found that a full packet of liquid droplets dissolved into discharge plasma resulted in more accurate spectroscopic results than fragmentary vaporization of droplets [36]. It has been shown that droplet dynamics are governed by several forces, such as electrical, aerodynamic, surface tension, internal viscous, and gravity during discharge plasma downward drag [23]. According to recent research investigations [19, 24, 29, 43, 56], homogeneous deposition coatings can be achieved by dissolving liquid droplets in atmospheric pressure plasmas at various precursor and gas flow rates.

2. Charging of liquid droplets in atmospheric pressure plasma

To understand the charging of liquid droplets in APP, we consider a fluid model in spherical coordinates in one dimension for a recombination-dominated equilibrium helium discharge plasma at atmospheric pressure. The radius of the sphere is (r), and it measures the distance from the symmetry axis. The initial densities of electrons and ions are assumed as ${n_e}^{(0)}$ and ${n_+}^{(0)}$, respectively, while the normalized fluid model is able to predict and identify the dynamic interaction among the droplets and atmospheric pressure plasma. Suppose, a spherical liquid droplet of radius (r_d) ($r_d \ll$ Debye length (λ_D)) is immersed in the discharge plasma and the recombination starts occurring on the surface of droplet after charging. As a result, the negative potential is acquired on the surface of droplet due to the higher thermal flux of electrons than ionic species to balance the flow of electrons and ions around the droplet [1, 2]. The growth of this process can be demonstrated in three possible steps (I, II, and III) as exhibited in **Figure 1**. The surface of liquid droplet is neutral before entering into discharge plasma as shown by step I. When the droplet interacts with atmospheric pressure plasma, the electrons rushed toward the droplet due to high thermal speed and formed a layer around the surface of the droplet at step II, which transform a neutral droplet into completely negatively charged surface of the droplet. This shows that the electrons develop a strong veneer on the surface of droplet and ultimately form a sheath around the droplet as expressed in step III for the single isolated droplet in the stationary discharge plasma. For simplicity, we consider direct ionization and dissociative recombination of helium ions and electrons for the numerical modeling.

In equilibrium, $k_i n_{He} n_e - k_R n_e n_+ = 0$, where k_i is the ionization rate of reaction, k_R is the recombination rate of reaction in the helium plasma, and n_e and n_+ are the densities of electrons and ions, while n_{He} is the neutral helium gas density. We assume a constant temperature of electrons to sustain the equilibrium in this 1D model,

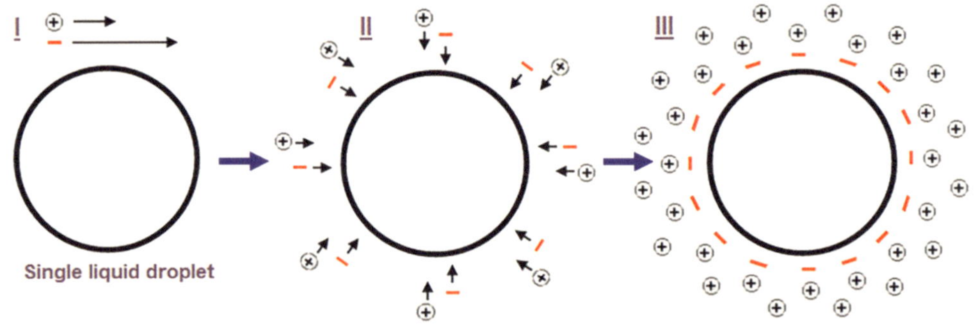

Figure 1.
Development of floating potential on the surface of droplets in APP.

although this assumption will require a further investigation in the future. Since the discharge plasma is assumed as quasi-neutral ($n_+^{(0)} = n_e^{(0)}$), the Boltzmann distribution is considered as suitable to describe the electrons in APP, and mathematically it can be written as $n_e = n_+^{(0)} \exp\left({}^{e\Phi}/_{kT_e}\right)$, where kT_{e} is the electron temperature expressed in eV and Φ is the electrostatic potential. We assumed an equilibrium state where the ionization and recombination rates in the discharge plasma were in balance. Further, the description of one-dimensional normalized fluid model equations for the discharge plasma is available in [37], which are used for the plasma modeling in the current scenario.

2.1 Effect of perturbation

The occurrence of perturbation is essential after the introduction of nebulizer droplets in the atmospheric pressure plasma, while the role of perturbation can be expressed by using a power series method [38]. We applied a classic technique to explore the behavior of liquid droplet using power series as $n = n + \alpha^1 n^{(1)} + \alpha^2 n^{(2)} + \ldots$. This approach is applicable for simple cases and becomes more complex at higher orders of magnitude. For lowest order, the density and velocity of discharge species after simplifications can be written as:

$$n^{(0)} \approx 1 + \eta_0 \frac{e^{-\rho}}{\rho} \tag{1}$$

$$v^{(0)} \approx \eta_0 e^{-\rho} \left(\frac{1}{\rho} + \frac{1}{\rho^2} \right) \tag{2}$$

where ρ is the normalized radius of droplet and $n_0 = 1 + \eta_0$, and η_0 is the arbitary constant. The above system of equations is reformulated by letting $\tau -> 0$ and after simplication, the solution of these equations is discussed around and at the surface of droplet in APP.

When the fluid model equations in normalized form are solved numerically under the specific boundary conditions in the discharge plasma, the parameters, such as floating potential, ionic species flux, the electric field, and ionic species density, are evaluated to describe the characteristics of APP in the presence of water droplets.

The electrons in discharge plasma are responded rapidly than heavy ionic species when the liquid droplet enters from the nebulizer. These electrons immediately interact with the liquid droplet and accumulate as a layer on the surface of droplet due to their higher thermal flux than ions [39, 45]. The APP is positively charged as compared to the deposition of negative charge on the surface of droplet due to the accumulation of electrons. The negative potential on the surface of the droplet is screened because Debye shielding confines the potential variations in this particular small region, called as a thin layer. The formation of a thin layer is termed as a sheath, and it forms a potential barrier to extinguish further spread of negative charge. The distribution of potential is progressed around the droplet when the dimensionless radius of discharge plasma channel alters from 0.01 to 0.1 as shown in **Figure 2(a)**. The sharply rising edge in electric potential around the droplet exhibits that the floating potential is immediately strengthened after entrance into atmospheric pressure plasma because of charging. Eventually, this develops a balance of electronic and ionic fluxes around the droplet, and this is verified by the presence of constant flux in the inner region, when the normalized radius (ρ) varies from 0.01 to 0.1 approximately as appeared in **Figure 2(b)**. It exhibits that the dynamic behavior of discharge participating species is evolved in the mentioned domain of normalized radii of droplet from 0.01 to 10. Therefore, it is evident from **Figure 2** that the characteristics of charge carriers are changed in the presence of nebulizer droplets and effectively perform an important role for the charging of the droplet. Based on this analysis, it is concluded that system of fluid model equations becomes simplified in the inner region due to the development of constant flux.

To distinguish the attributes of charge carriers around and on the surface of the droplet, the distributions of drift velocity and electric field are compared as shown in **Figure 3**. Considering the flux balance conditions around the droplet, the strength of magnitudes of drift velocity and electric field is altered from higher to lower values. As the intense charged layer is formed at the surface of the droplet, which is responsible for the organization of distorted electric field at the surface of the droplet. It is apparent from **Figure 3(a)** that the lowering in the drift velocity and electric field is linearly varying from $\rho = 0.01$ to 1, while a sharp fall appears at higher normalized radii ρ. This shows that the magnitudes of drift velocity and electric field are prominently reduced to very small values away from the surface of the droplet. This situation is entirely transformed at the surface of droplet, when the velocity of ionic species and strength of electric field are certainly boosted to the higher values as a function of droplet radius as exhibited in **Figure 3(b)**. The comparison of both situations reveals that the tremendous modification occurs at the surface of droplet

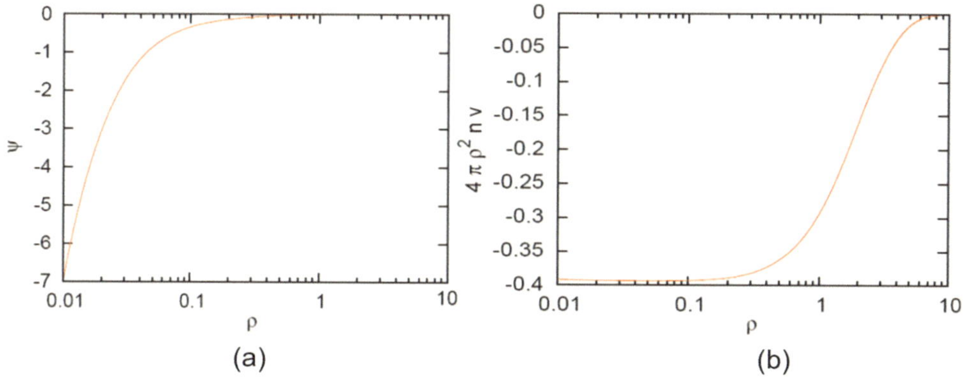

Figure 2.
(a, b) Spatial profiles of normalized floating potential (ψ) and ionic species flux around the droplet in APP.

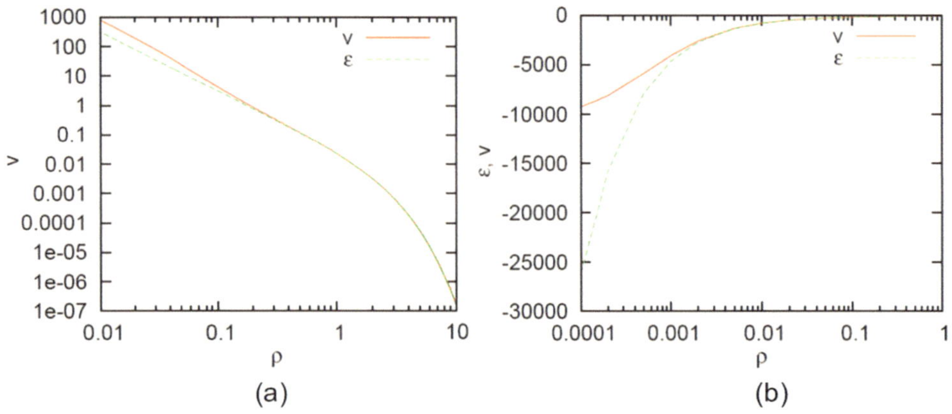

Figure 3.
(a, b) Spatial distributions of normalized drift velocity and electric field around and at the surface of droplet in APP.

due to the formation of sheath including the accumulation of electrons. This can be verified from the sharp and steady change in the drift velocity and electric field as viewed from **Figure 3**. The sharp distortion in electric field from 0.0001 to 0.001 exhibits that the sheath developed around the droplet due to the deposition of negative charge on the surface of the droplet.

Figure 4(a) shows that the normalized magnitude of the droplet floating potential at the surface continuously changes over time due to the strong perturbations caused near the surface by ions and electrons. It is also evident from the mathematical expressions in Eqs (1) and (2) that the spread of perturbation is uniformly homogeneous around the droplet. To explore the interaction of droplets during the transport in the plasma channel, the surface charge on the droplet is evaluated from the floating potential ($\psi 4\pi\varepsilon_0 r_d$) [40] as well as from the Rayleigh method ($8\pi\sqrt{\varepsilon_0 \gamma r_d^3}$) [41]. Based on **Figure 4(b)**, it can be demonstrated that the surface charge of the droplet is appropriately less than the charge determined by Rayleigh's criterion. In comparison with droplet charging, droplets remain intact during residence time and are not decomposed into smaller particles. Therefore, the scenario of droplet charging can be supported. The characteristics of floating potential are distinguished

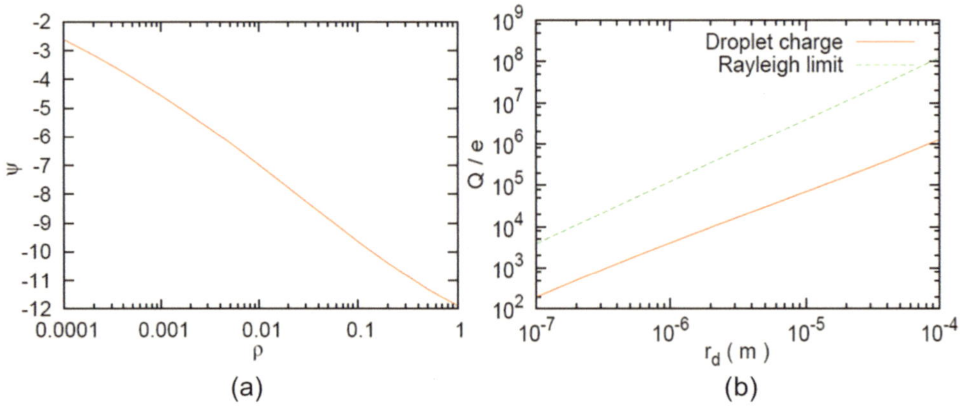

Figure 4.
(a, b) Distribution of normalized floating potential at the surface of droplet and charge ratio (Q/e).

from **Figures 2(a)** and **4(a)** showing the strong behavior around and on the surface of the droplet. This shows that the evaporation is not a very important mechanism in certain domain of droplet radii (>5 μm), and the droplets survive successfully during the passage in the plasma channel, which act as the seed for nonuniform surface depositions. Therefore, the behavior of droplets in helium APP can be described by the spatio-temporal characteristics using the nonequilibrium Langumir-Knudsen law [42] and normalized one-dimensional normalized fluid model [37] in spherical coordinates.

3. Multidimensional coupled fluid-droplet model for PlasmaStream system

Now the multidimensional modeling approach is preferred to explain and interpret the liquid-plasma interaction, while the multiphase flow can be described by the Eulerian–Lagrangian numerical scheme. In this approach, the Lagrangian methods are used to investigate droplet dynamics, and Eulerian methods are used to investigate the kinetics of the He-N_2 gas mixture, in addition to major collision interactions initiated by chemical reactions. As a result of adding the coupling terms to the conservation equations of mass, momentum, and energy of the gas and liquid phases [43, 44], the fluid model of the plasma gas and stochastic liquid droplet model of the liquid precursor droplets were integrated including mutual interactions. The following set of equations is used for the modeling of flowing gas mixture as:

$$\frac{\partial \rho_g}{\partial t_1} + \rho_g \vec{\nabla}.\vec{u}(r,z,t) = S \tag{3}$$

$$\frac{\partial\left(\rho_g \vec{u}(r,z,t)\right)}{\partial t_1} = \vec{\nabla}P + \vec{\nabla}.\vec{\Gamma}_\mu \nabla \vec{u}(r,z,t) + S_M \tag{4}$$

$$\frac{\partial\left(\rho_g h\right)}{\partial t_1} = \vec{\nabla}.\frac{k}{C_p}\vec{\nabla}h + \vec{j}.\vec{E} + S_E \tag{5}$$

$$\rho_g \frac{\partial Y}{\partial t_1} = \vec{\nabla}.D_{AB}\nabla Y + S \tag{6}$$

Table 1 lists the terminology and description of variables for the above set of equations. Eqs (3–6) represent the fluid model equations of mass density, momentum, and energy of the gas mixture, while the models are coupled through source terms (S, S_M, and S_E). In this case, Implicit Continuous Eulerian (ICE) method is used because this has already been tested for the numerical solution of gas flows of any Mach number [33]. This has also considered as the effective for the calculation of terms containing momentum exchange between two phases. For the presence of species in APP, the kinetics behavior of discharge species (e^-, He^+, He_2^+, N_2^+, He^*, and He_2^*) are determined by solving the generalized continuity equation as written as follows:

$$\frac{\partial n_{sp}(r,z,t)}{\partial t_2} + \vec{\nabla}.\left(\vec{\Gamma}_{sp}(r,z,t) + n_{sp}\vec{u}(r,z,t)\right) = S_{sp} \tag{7}$$

where the index **sp** corresponds to the distinctive species (electrons, ions, neutrals, and radicals), $\vec{\Gamma}_{sp}(r,z,t) = \text{sgn}\left(q_{sp}\right)\mu_{sp}\vec{E}n_{sp} - D_{sp}\vec{\nabla}n_{sp}$ indicates the flux of discharge

Symbols	Physical description
ρ_g, $\vec{u}$, P, Γ_μ, k, $\vec{j}$, $\vec{E}$, h, Y, D_{AB} and S	Density of gas mixture, velocity vector in the radial and axial directions of gas (u_r, u_z), operating pressure, momentum, thermal conductivity, joule heating, electric field in the radial and axial directions (E_r, E_z), diffusion coefficient, enthalpy, vapor species mass fraction, and various source terms in the set of Eqs (3–6).
r_d	Radius of droplet
λ_D	Debye length
C_P	Specific heat at constant pressure
$\vec{x}_P$	Position of droplets in the parcels
$\vec{u}_P$	Velocity of droplets in the parcels
$\vec{u}'$	Velocity of gas with fluctuations
ρ_l	Density of liquid precursors
k_B	Boltzmann constant
g_z	Gravitational constant
R	Universal gas constant
$c_D = (24/R_e)(1 + R_e{}^{2/3}/6)$	c_D is used in the calculation of drag coefficient
$R_e = \left(2\rho_g\left\|\vec{u} + \vec{u}' - \vec{u}_d\right\|r_d\right)/\mu_g(\overline{T}_g)$	Reynolds number of droplets
$\vec{F} = K_P\left(\vec{u} + \vec{u}' - \vec{u}_d\right) - (1/\rho_l)\vec{\nabla}p + mg_z\hat{k} + q\vec{E}$	Forces acting on the droplets
$\mu_g(\overline{T}_g)$	Viscosity of gas
$K_P = \frac{3}{8}\frac{\rho_g}{\rho_l}\frac{\left\|\vec{u}+\vec{u}'-\vec{u}_d\right\|}{r_d}c_D$	Represents the evaluation of a variable
$\overline{T}_g = \frac{(T_g+2T_d)}{3}$	$\overline{T}_g$ and T_g are the average and local gas temperatures, respectively
$Nu_{(g,\ l)}$	Nusselt number of gas and liquid
Y_{V_s}	Liquid vapor mass fraction on the surface of the droplet
Y_V	Liquid vapor mass fraction
λ_g	Heat conductivity of gas
$Pr = \mu_g c_{pg} / \lambda_g$	Gas phase Prandtl number
c_l	Specific heat of liquid at constant pressure
T_s	Surface temperature of droplet
$P_g = \rho_g R T_g\ (Y_v / W_v + Y_I / W_I)$	Equation of state
$h_g = (c_{pv}\ Y_v + c_{pI}\ Y_I)\ T_g = c_{pg} T_g$	Equation of state
W_V, W_I	Molecular weight of liquid vapors and gas
c_{P_V}, c_{P_I}	Specific heat of vapor species and inert gas species at constant pressure
W_e	Weber number
$h_l(T_d) = C_{pv}\ T_d - L(T_d)$	Enthalpy of liquid droplets

Symbols	Physical description
$f'(r_d) = (1/\bar{r}_d)\exp(-r_d/\bar{r}_d)$	X-squared distribution function, r_d and $\bar{r}_d$, respectively, are the radius and number-averaged droplet radius for initial distribution
μ_{sp}, T_{sp} and q_{sp}	Mobility, temperature, and charge for discharge species of discharge plasma
$S = -\int 4\pi r_d^2 \dot{r}_d \rho_d f' d\vec{u}_d dr_d dT_d$	Source term for mass transfer
$\vec{S}_M = -\int f'\rho_d\left(\frac{4}{3}\pi r_d^3 \vec{F} + 4\pi r_d^2 \dot{r}_d \vec{u}_d\right) d\vec{u}_d dr_d dT_d$	Source term for axial momentum transfer
$S_E = -\int f'\rho_d \left\{ \begin{matrix} 4\pi r_d^2 \dot{r}_d h_l(T_d) \\ +\frac{4}{3}\pi r_d^3 \left\{ \begin{matrix} c_l \dot{T}_d \\ +\vec{F}.\left(\vec{u}_d - \vec{u}\right) \end{matrix} \right\} \end{matrix} \right\} d\vec{u}_d dr_d dT_d$	Source term for gas enthalpy

Table 1.
Variables in fluid-droplet model (2D and 3D).

species including drift and diffusion components, μ_{sp} and D_{sp} correspond to mobility and diffusion coefficient of discharge species, respectively, while the Einstein relation ($D_{sp} = \mu_{sp} k_B T_{sp}/q_{sp}$) is used for the calculation of numeric values. The discharge species density is denoted by n_{sp}, and S_{sp} exhibits to the source term for the construction and destruction of discharge species based on the chemical reactions as investigated in [45]. The electron mean energy plays a vital role in the transport coefficient and reaction rate coefficient of electron impact reactions that are updated at every time step in the plasma model indicating a strong coupling between the two-phase flow. The electron mean energy (ε) is determined by using the electron energy density balance equation as:

$$\frac{\partial n_\varepsilon(r,z,t)}{\partial t_2} + \vec{\nabla}.\left(\vec{\Gamma}_\varepsilon(r,z,t) + \varepsilon n_e \vec{u}(r,z,t)\right) = S_\varepsilon \tag{8}$$

The electron energy density is calculated using $n_\varepsilon(r,z,t) = n_e(r,z,t)\varepsilon(r,z,t)$, where n_e exhibits the electron density and $\vec{\Gamma}_\varepsilon$ corresponds to the flux. Considering all of the terms in [46] together, the impact of joule heating, electron energy losses, and implicit correction terms is denoted by the energy source term (S_ε). To solve the particle balance equation and electron energy density equation numerically, the alternating direction implicit (ADI) solver is used [47]. For the estimation of the electric field ($\vec{E}$), Poisson's equation is coupled with the plasma fluid model, and it can be expressed as:

$$\vec{\nabla}.\left(\varepsilon_0 \vec{E}(r,z,t)\right) = -\sum_{sp} q_{sp} n_{sp}(r,z,t) \tag{9}$$

where ε_0 is the permittivity of the free space, and the space charge density is the product of species charge (q_{sp}) and density (n_{sp}). From the gradient of the electric potential ($-\partial V/\partial r$, $-\partial V/\partial z$) in two dimensions, we calculate the electric field strengths along r and z. Here, the numerical solution to Poisson's equation is efficiently achieved by using the successive over relaxation (SOR) method as applied in [48]. The local field approximation is used to calculate the effective electric field for the transport of ions in discharge plasma, in which the transport and reaction rate coefficients are tied to the electric field [49]. In this numerical model, secondary emission of electrons is not involved because ions have small energies compared to electrons, according to [50].

3.1 Geometry of a plasma chamber and experimental chamber

A range of siloxane monomers including hexamethyldisiloxane (HMDSO), *n*-hexane, tetraethyl orthosilicate (TEOS), and water were used to deposition coating on a PlasmaStream system using the laser diffraction particle size analysis technique as shown in **Figure 5(a)**. For plasma generation, a 7.5-cm-long polytetrafluoroethylene (PTFE) tube with a 1.5-cm inner diameter is used. **Figure 5(b)** shows the schematic geometry for the simulation region used in modeling the experiment, in which an oscillating sinusoidal potential (V) is imposed to a thin metal electrode at a specific frequency (f), and the substrate is grounded. Experimental investigations are performed on silicon substrates with resistivities ranging from 0 to 100 V cm^{-1} (450 and 300 mm thick), for the deposition of coatings [19]. Assuming axisymmetry at the inlet boundary, we introduce the laminar profile of gas flow velocity. **Figure 5(b)** shows a pin metal electrode separated by two gas flows (Q_1 and Q_2 l min^{-1}) on the top side of the chamber (AB). The nebulizer droplets are introduced from the central location (B) with a flow rate Q_3 μl min^{-1} and are mixed with APP. At the central channel of APP, the droplets of different types of precursors which are served as discrete parcels are propelled along a stream of He-N_2 (99% He and 1% N_2) carrier gas until they reach the gaseous state (**Table 2**).

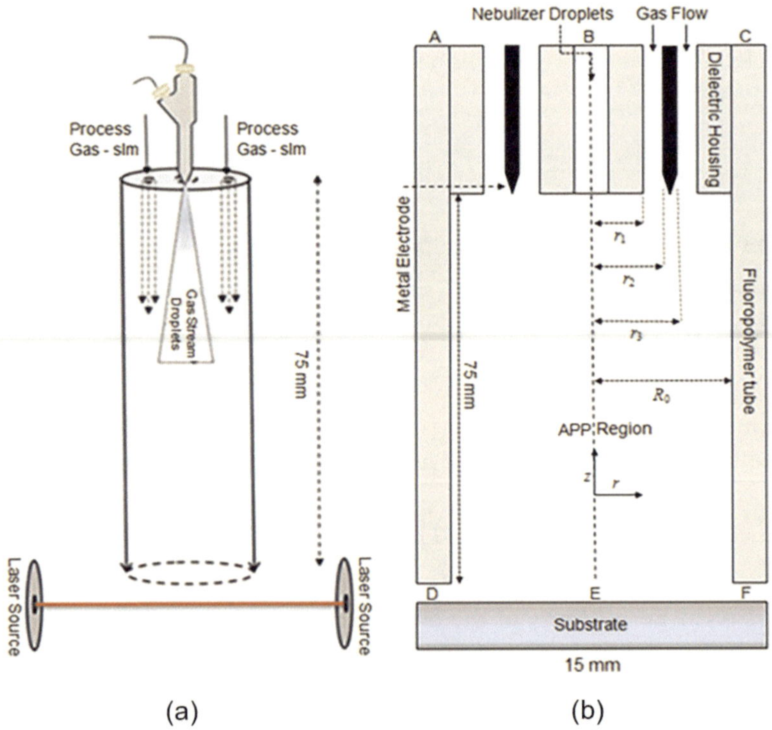

Figure 5.
(a) Symbolic diagram of experimental chamber of PlasmaStream™ atmospheric pressure plasma jet deposition system with laser source [19] and (b) schematic geometry for the fluid-droplet model.

Inlet (z = 0)	**Exit (z = L)**
$u = 0$ for for $0 < r \leq r_1$	$\partial(\rho u)/\partial z = 0$
$u = Q_1/\pi(r_2^2 - r_1^2)$ for for $r_1 < r \leq r_2$	$\partial h/\partial z = 0$
$u = 0$ for $r_2 < r \leq r_3$	$\partial Y/\partial z = 0$
$u = Q_2/\pi(R_0^2 - r_3^2)$ for for $r_3 < r \leq R_0$	$\partial v/\partial z = 0$
$v = 0, w$ = User's choice, $Y = 0$	
$V = V_0 \sin(2\pi ft)$ for $r_2 < r \leq r_3$	V = Grounded
Centerline (r = 0)	**Glass wall (z = R_0)**
$v = 0, w = 0$	$v = 0, w = 0$
$\partial h/\partial r = 0$	$\partial h/\partial r = 0$
$\partial Y/\partial r = 0$	$\partial Y/\partial r = 0$
$\partial u/\partial r = 0$	$u = 0$
$\partial V/\partial r = 0$	$\partial V/\partial r = 0$

Table 2.
Fluid-droplet model (2D) using initial and boundary conditions.

3.2 Stochastic liquid droplet model

For a single liquid droplet falling vertically under the combined influence of a number of forces, the following equations described in [17] estimate the transient position of the droplet, its velocity, radius, and temperature.

$$\frac{d\vec{x}_p}{dt_1} = \vec{u}_p \tag{10}$$

$$\frac{d\vec{u}_p}{dt_1} = \vec{F} \tag{11}$$

$$\frac{dr_p}{dt_1} = \left(\frac{\lambda_g}{\rho_1 c_{p_g}}\right)\left(\frac{Nu_g}{2r_p}\right)\left(\frac{Y_{V_s} - Y_V}{1 - Y_{V_s}}\right) \tag{12}$$

$$\frac{dT_p}{dt_1} = \left(\left(\frac{3\lambda_1 Nu_1}{2\rho_1 c_1 r_p^2}\right) + 3\frac{dr_p/dt_1}{r_p}\right)(T_s - T_p) \tag{13}$$

where the index p corresponds to a particular parcel and the details of variables in Eqs (10–13) are available in **Table 1**. In this model, an x-squared distribution $\left(R(r_d) = (1/\overline{r}_d)e^{-\frac{r_d}{\overline{r}_d}}\right)$ of droplets in various parcels is introduced by taking the rectangular computational mesh, where r_d and $\overline{r}_d$ are the radius and mean radius of droplets. The index i and j of each cell increases along the radial and axial directions. The accuracy of the numerical simulations is obtained by mesh independence, and the mesh size is implemented as $(r, z) = (20, 30)$ in these simulations, which are similar as used in [17, 21]. In the above equations, the dynamic nature of droplets is similar in each parcel, but collisions occur between different parcels. The formula for the evaluation of the collision frequency (ν_c) is mentioned as follows:

$$\nu_c = n_{p_2}\pi(r_{col} + r_{con})^2\left|\vec{u}_r\right|/V_{ijk} \tag{14}$$

In the above Eq. (14), the collector and contributor droplets are shown by r_{col} and r_{con}, while the relative velocity between parcels is denoted by $\vec{u}_r = \vec{u}_{p_1} - \vec{u}_{p_2}$. The number of droplets in the associated parcel is exhibited by n_{p_2}, and the volume of the cell containing both parcels is represented by V_{ijk}. $P_0 = e^{-\overline{n}}$ is employed to calculate the probability of absent collision. Here, $\overline{n} = \nu_c\Delta t_1$ is the mean number of parcels and Δt_1 is used the computational time step in the case of Lagrangian approach. The collision and critical impact parameters (b, b_{cr}) are described below to express the collisions between the droplets by the following relations as:

$$\left.\begin{aligned} b &= (r_{col} + r_{con})\sqrt{Y} \\ b_{cr} &= (r_{col} + r_{con})\sqrt{E_{coal}} \end{aligned}\right\} \tag{15}$$

where E_{coal} is the coalescence efficiency and defined as $E_{coal} = \min\left(1.0, 2.4\zeta(\gamma)/W_e\right)$. The complex function $\xi\ (\gamma)$, is approximated by the polynomial for simplicity [17] as $\xi\ (\gamma) = \gamma^3 - 2.4\ \gamma^2 - \gamma$ and $\gamma = r_{con}/r_{col}$. The parameter (Y) is determined by a random number that exists within the interval (0, 1). In case the condition $(b \geq b_{cr})$ is true, then the result of a collision is grazing, which arises within a brief period following an injection of droplet pulse. Collisions caused by grazing occur when the droplets in the parcels and collectors collide so that they preserve their size radii and temperatures, but this collision also changes their velocities. The droplets in the plasma chamber coalesce when $(b < b_{cr})$ occurs during downward displacement, leading to the formation of a new droplet with a higher radius. Nevertheless, the main step is to combine stochastic parcel method and ICE technique, in which gas flow rates and droplet velocities are integrated. For the discretization of mathematical equations, it is recommended to use first-order accuracy of finite difference method (FDM). A detailed description of the discretization procedure for these models is given in [17, 46]. Accordingly, **Tables 3–5** contain the description of the fluid-droplet model, including parameters and constants from an online database [51].

Fluid-droplet model variables	**Initial input values**
Density of quasi-neutral plasma at t = 0.0	1.0×10^9 cm^{-3}
Axial length and diameter of chamber	7.5 and 1.5 cm
External imposed potential (V_0)	–13.5 kV
Driving frequency (f)	20 kHz
Operating pressure (p)	760 Torr
Initial gas temperature (T_g)	300 K
Precursor droplet temperature (T_d)	292 K
Injection droplet velocities (u_r and u_z and u_θ)	$(1.5 \times 10^3, 1.0 \times 10^{-3}$ and $1.0 \times 10^{-3})$ cm s^{-1}
Gas species density (ρ_g)	0.1617×10^{-3} g cm^{-3}
Vapour species mass fraction (Y)	0.01
Swirl velocity (w) of gas	1.0×10^{-3} cm s^{-1}

Fluid-droplet model variables	Initial input values
Diffusivity of helium gas	1.6313 $cm^2\ s^{-1}$
Time steps for droplets (Δt_1) and plasma (Δt_2)	0.1×10^{-5} and 1.0×10^{-9} s
Initial energy of electrons and ions	0.5 and 0.026 eV

Table 3.
Fluid-droplet model initialization.

Liquid precursors	Surface tension × 10^3 (dyne cm^{-1})	Dynamic viscosity (dyne s)	Boiling point (K)	Mass density (10^{-3} × g cm^{-3})
HMDSO ($C_6H_8OSi_2$)	1.57	4.88×10^{-3}	374.00	764.00
n-Hexane (C_6H_{14})	1.84	3.00×10^{-3}	341.88	660.30
TEOS ($C_8H_{20}O_4Si$)	2.28	7.90×10^{0}	441.10	934.00
Water (H_2O)	7.28	8.90×10^{-3}	373.13	1000.0

Table 4.
Parameters used in fluid-droplet model (2D).

Time/density	t_1 = 0.6 ms	t_2 = 1.6 ms	t_3 = 2.6 ms	Precursor
Electron density (cm^{-3})	1.05×10^{11}	1.25×10^{11}	1.50×10^{11}	TEOS
Electron density (cm^{-3})	4.10×10^{11}	4.20×10^{11}	4.12×10^{11}	HMDSO
Vapor species density (g cm^{-3})	4.10×10^{-5}	4.70×10^{-5}	5.30×10^{-5}	TEOS
Vapor species density (g cm^{-3})	1.15×10^{-4}	1.25×10^{-4}	1.13×10^{-4}	HMDSO

Table 5.
Density of electrons and vapor species using three-dimensional fluid-droplet model (3D).

3.3 Initial and boundary conditions

Assumptions are made regarding laminar profiles of the axial, radial, and swirl velocities (u, v, w) in the PlasmaStream model. For grounded electrodes and solid metal substrates, the flux of electrons, heavy charged particles, neutral gas species, and electron energy density are calculated by considering the modified boundary conditions during the alternate cycles of sinusoidal voltage as mentioned in [52, 53]. The expression $v_{\mathrm{th,sp}} = \sqrt{8k_BT_{\mathrm{sp}}/\pi m_{\mathrm{sp}}}$ is used for the evaluation of thermal velocity. The freeware Boltzmann solver (BOLSIG+) [54] was employed for the estimation of electron mobilities as well as reaction rate coefficients of excitation and ionization processes. Moreover, two different time steps are implemented to control the liquid phase of droplets (Δt_1) and plasma in the fluid-droplet model (Δt_2). Plasma equations can be numerically solved with a small time step constraint imposed on the plasma model. Courant–Friedrichs-Levy (CFL) condition is met when time step is selected for the case plasma model using an inequality, $\Delta t_2 < \left(\varepsilon_0/\sum_{\mathrm{sp}}|q_{\mathrm{sp}}|\mu_{\mathrm{sp}}n_{\mathrm{sp}}\right)$ [37] as compared to the liquid droplet phase. On the other hand, the coupling source terms are used to solve the numerical solution of the liquid and gas phases by using the average values of electric field. For the gas and precursor droplets, we take into account small values of swirl velocities, since they have no significant effect on the numerical solution of two-dimensional equations. **Table 3** summarizes the operating conditions for the fluid-droplet model as described in the experiment.

4. Dynamics of interaction in two-phase flow

Utilizing a coupled fluid-droplet model, this section investigates the dynamics of interactions between different precursor droplets and APP. By displaying a quantitative understanding of droplet transit in the plasma chamber, we are able to explore the temporal mean profiles of droplet radii and temperatures, droplet count and size distributions, and vapor species density. To establish the authenticity of modeling outcomes, we performed a comparison of simulation results with experimentally measured size of HMDSO droplets using the laser diffraction particle size analysis technique in which the plasma discharge tube was positioned between the laser source and the capture lens as rationalized in detail in [19]. In order to evaluate the effectiveness of the results derived using the fluid-droplet model, the effect of evaporation is compared by taking four liquid precursors into account: HMDSO, n-hexane, TEOS, and water. The time for the evaporation of the entire pulse of droplets is dependent on many factors, while we address an impact of precursor and gas flow rates to describe the characteristics of APP. A similar profile of gas velocity is assumed as provided by the experimental measurements for the numerical simulations [19]. The initial injection velocity of droplets is distinctive in various sections of this chapter.

4.1 Behavior of two-phase flow

Immediately after the liquid droplets interact with discharge plasma, the evaporation of the droplets begins. This occurs due to multiple forces that are simultaneously applied to the liquid droplets, such as electrical ($\vec{F} = q\vec{E}$), aerodynamic drag ($\vec{F} = K_p\left(\vec{u} - \vec{u}' - \vec{u}_d\right) - (1/\rho_l)\vec{\nabla}p$), internal viscous ($\mu_v$), surface tension, and gravitational ($\vec{F} = m\vec{g}_z$) during this interactive process in the PlasmaStream jet deposition system. The effect of interaction between two-phase flow is investigated at different time instants, such as 3, 6, and 9 ms as shown in **Figure 6**. Because the volatile HMDSO liquid droplets are exposed to atmospheric pressure plasmas within discharge plasma pulses, the vaporization rate of liquid droplets increases rapidly from 0 to 3 ms, which can be proved by a sharp peak in vapor species density along the pulse of droplets. In discharge plasma, the impact of droplet vaporization increases the lifetime of remaining droplets. The lifetime of the survived droplets in discharge plasma is increased by the impact of vaporization of droplets that further decelerates the evaporation of the remaining droplets in the plasma chamber. This effect expands spatially with time because the vapor species density ($\rho_g Y$) increases along the evaporating pulse of droplets as shown in **Figure 6(a–c)**. Due to the higher temperatures of the plasma gas mixture in comparison with the liquid droplets, vaporization of droplets dominates in nonequilibrium APP. Based on the increase in vapor species density, the droplets vaporize fairly rapidly between 0 and 6 ms; however, this process impels to slow down with a reduction in the number of droplets in the APP at 9 ms, as discussed in [43].

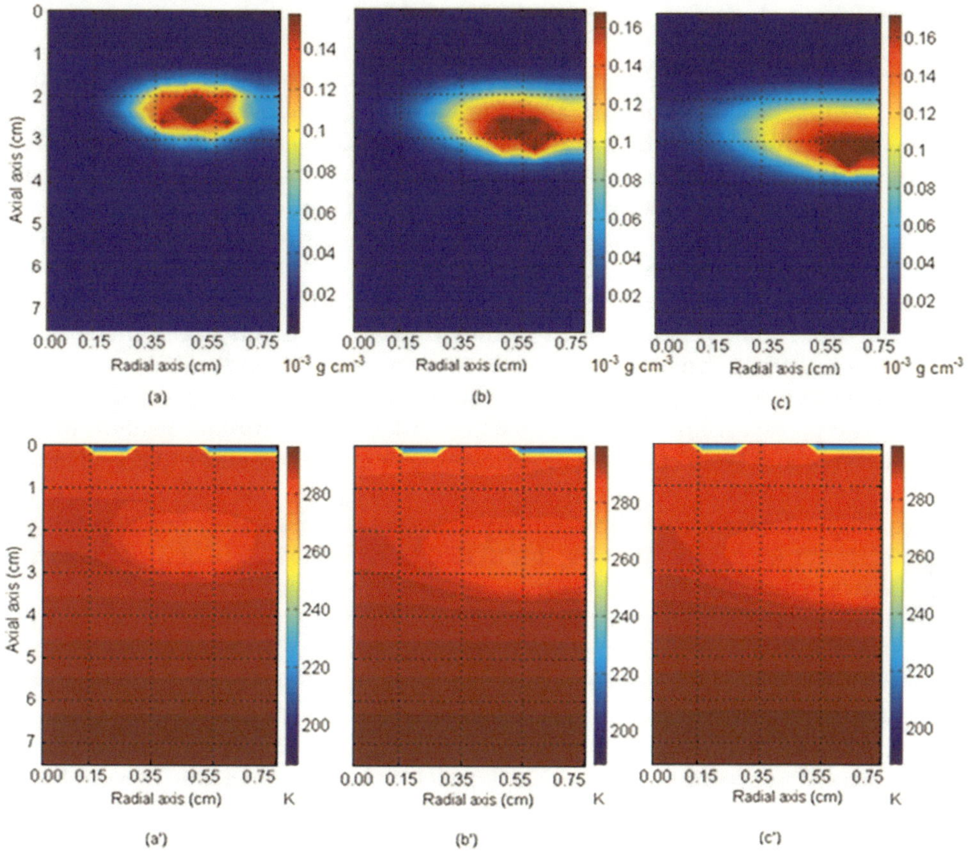

Figure 6.
Top row (a, b, c) shows vapor species density and bottom row (a′, b′, c′) exhibits temperature of the gas at 3, 6, and 9 ms using HMDSO precursor flow rate = 100 μl min^{-1}, gas flow rate = 5 l min^{-1}, f = 20 kHz, and V_o = −13.5 kV.

A strong gradient along the pulse of evaporating droplets is evident in the spatial distributions of gas temperature ($\overline{T}_g$) as illustrated in **Figure 6(a′–c′)**. Vapor species density is produced as an additional species that is incorporated after droplets transfer into vapor after phase shifting. Due to evaporation, the temperature of the gas surrounding the HMDSO droplets has dropped sharply. As can be seen from the spatial distributions of gas temperatures, the area around the droplets pulse has been significantly affected by the evaporation as they fall downward. In the plasma chamber, as indicated clearly at three distinct points in time (3, 6, and 9 ms), the temperature of gas falls as the droplets evaporate, but the impact of the vaporization increases. Consequently, this can result in a lowered gas temperature which is useful and effective for the sensitive biomedical and industrial applications in different operating conditions of gas flow and precursor flow rates. Observations of gas temperature profiles indicate that distortions, caused by droplet dispersal, contribute significantly to the formation of a cold plasma at atmospheric pressure. Therefore, the diminish in gas temperature along with the increase in species density endorses the turbulent nature of the liquid precursor of the HMDSO in He-N_2 atmospheric pressure plasma as shown in **Figure 6**.

4.2 Temporal evolution of a pulse of HMDSO droplets

We analyzed the temporal mean profiles of radius, axis velocity, and temperature of HMDSO droplets in APP in order to understand the role of evaporation in APP. **Figure** 7 shows the mean density of the vapor species during the entire lifetime of a pulse of droplets in the plasma chamber. A higher temperature of the APP enables the droplets to gain energy, resulting in evaporation around the surface of the droplets. While both the number of droplets and the temperature of the parcels are unchanged at the start, their radii and temperatures are reduced due to strong evaporation in domain I. Droplets in parcels contract in size due to evaporation, and some of them acquire nanometer-sized domain radii. For the various kinds of precursors in gas mixtures, the precedent for vaporization of droplets has been examined [17], and a similar benchmark is employed when trying to explain the evaporation behavior of droplets in APP. During interaction, a parcel is removed from the parcel distribution if its entire bundle of droplets evaporates. Droplet counts in parcels evaporate continuously, and its overall effect can be observed in the distribution of their mean radii. Due to drop-down of average temperature, the temperature of each droplet amplifies after gaining energy from the discharge plasma; however, due to the strong vaporization of droplets at the start, the rise in temperature of droplets diminishes in each parcel. Thus, the mean temperature of the remaining parcels actively curtails during the initial period of 1 ms. This is illustrated by a red dotted line in **Figure** 7 which indicates that the mean concentration of vapor species increases moderately. This occurs because of evaporation of droplets in the domain I rather than being converted to vapor.

In domain II, the droplets start enhancing in temperature as they attain partial steady states in the plasma chamber, as indicated by a black dashed line. Droplet-plasma interactions ultimately result in vapor species being transferred into discharge plasma, and this in turn drives a greater density of He-N_2 gas mixture. As a result, when intensive evaporation of droplets occurs, their momentary temperature quenches while in the case of reduced evaporation, it starts to increase, as demonstrated by the green dashed line in **Figure** 7 for domains I and II. It is seen at 8 ms as highlighted in the

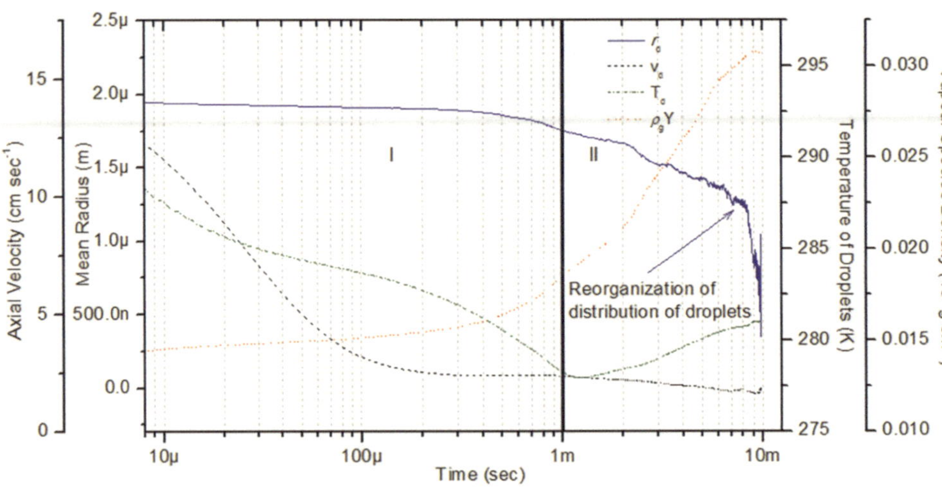

Figure 7.
Mean distributions of droplet radius (solid line), droplet axial velocity (dashed line), droplet temperature (dash-dotted line), and mean vapor species density (dotted) of parcels using HMDSO precursors at flow rate = 100 μl min^{-1}, gas flow rate = 5 l min^{-1}, f = 20 kHz, and V_0 = −13.5 kV.

domain II with an arrow that the evaporation accelerates sharply because the number of liquid droplets becomes very small at this stage. Droplets are distributed in a new manner after parcels of droplets are eliminated by evaporation, as indicated by the small spikes in the profile of mean radius of droplets. According to **Figure 7**, it is obvious that the density of the mean vapor species grows until it reaches a stationary state just before the phase shift of the entire pulse of droplets as indicated by the dotted red line.

4.2.1 Effect of HMDSO, n*-hexane, TEOS, and water*

Through the incorporation of a relevant precursor, the characteristics of surface deposition coating can be adjusted, resulting in a discharge plasma that was altered fundamentally by evaporation and finite-shared interactions between the droplets. We consider three kinds of liquid precursor droplets, for example, HMDSO, *n*-hexane, TEOS, and water to analyze the properties of APP and the size radii of droplets exists in the range from 4 to 24 μm. As a result of strong intermolecular interactions within their molecules, the HMDSO liquid precursor evaporates completely within 4 ms, while the *n*-hexane, TEOS, and water droplets survive for longer. It suggests that HMDSO is more reactive and volatile than *n*-hexane, TEOS, and water under similar operating conditions when interacting with He-N_2 discharge plasma. By avoiding the areas near the cathodes and grounded electrodes at the maximum value of discharge current density after 2 ms, the spatial profiles of vapor species and electron density are exhibited in the center of the plasma chamber, as shown in **Figure 8**. In the analysis of the electronic and vapor species distributions, the central section of the plasma chamber is used due to dominant effect of evaporation in APP rather than the area near the electrodes since it has higher density peaks. As demonstrated in **Table 4**, all of these precursors have varying physical and chemical characteristics.

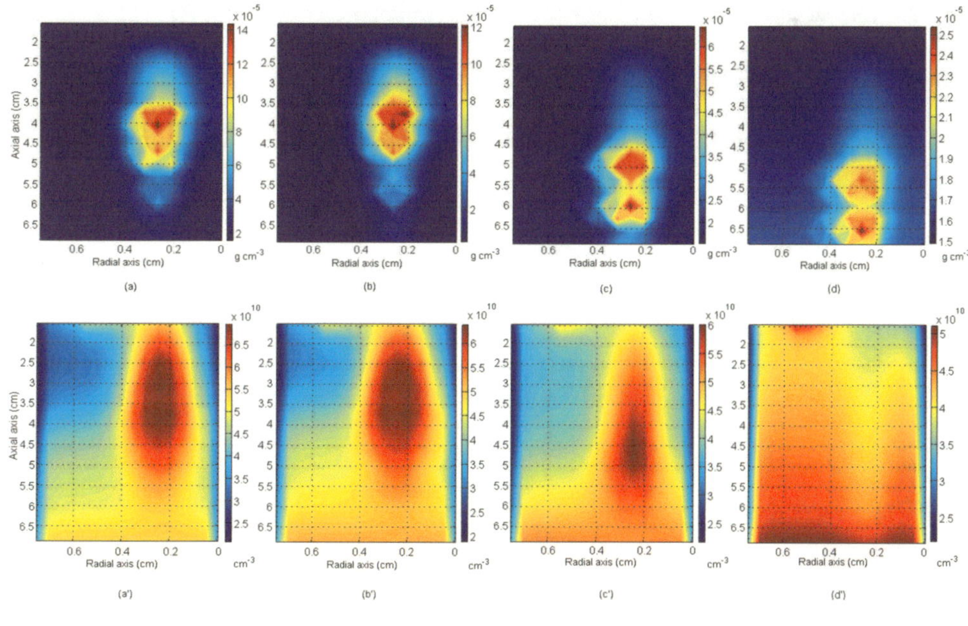

Figure 8.
Top row showed the spatial profiles of vapor species density and bottom row illustrated electron density, using (a, a′) HMDSO, (b, b′) n*-hexane (c, c′) TEOS, and (d, d′) water at precursor flow rate = 100 μl min^{-1}, gas flow rate = 5 l min^{-1}, f = 20 kHz, and V_o = −13.5 kV.*

Figure 8(a–d) reveals that HMDSO precursor has higher vapor species density than the three other precursors, including *n*-hexane, TEOS, and water because of the high level of evaporation. Several factors influence the rate of evaporation when droplets interact with discharge plasma, such as surface tension, dynamic viscosity, mass density, and boiling point of the precursor. Droplets' surface tension provides a strong resistive force against evaporation, and since HMDSO has a very small surface tension in contrast to other precursors, this implies rapid evaporation of the droplets (as shown in **Table 4**). The temperature of gas mixtures amplifies due to higher value of electron mean energy ($\varepsilon \sim 0.4$ eV) around the pulse of droplets than other discharge species in APP. A convection effect occurs during the active phase of the discharge current pulse in which the precursor droplets can vaporize as a result of the energy gain in the He-N_2 gas mixture. This happens due to the impact of electric field and neutral gas density which is evolved by the exchange of energies in two-phase flow through coupling source terms as represented in the group of Eqs (3–6). Since water droplets have a high mass density value, they are pulled downward by gravity to a much greater extent than liquid precursor droplets. According to **Figure 8(a–d)**, the evaporating pulse of water droplets is more likely to pull downward than other precursors. HMDSO and *n*-hexane liquid precursors do not exhibit significant spatial differences along the pulse due to their small differences in mass density. Hence, the vaporization of droplets causes changes in the density of the vapor species and neutral gas, influencing the discharge properties directly.

Table 6 shows that the numerical values of vapor density for different species diminish from HMDSO to water. Due to a high rate of evaporation, the gradient of electron density along the pulse of droplets in HMDSO and *n*-hexane precursors is relatively larger than that of TEOS and water precursors as shown in **Figure 8(a′–d′)**. This develops conductive pathways and can be grasped by HMDSO or *n*-hexane as active channels, rather than TEOS or water. Moreover, water (H_2O) is regarded as the best example of hydrogen bonding because it is polar molecule [55]. As well as having intensive intermolecular forces between their molecules, water droplets have a higher surface tension and density. These factors all contribute to the increased lifetime of water droplets in APP and a weak rate of evaporation of water relative to HMDSO, *n*-hexane, and TEOS precursors. So, there is a higher probability of smashing of water droplets on the substrate surface due to survival in the plasma and the survived droplets responsible to build a nonuniform surface deposition. Based on the numerical values of vapor species and electron densities for these precursors, it can be seen that the trend is reducing as exhibited in **Figure 8**. Compared to other precursors, the electron density gradient in water droplets is sluggish, which provides an explanation for small vaporization of water droplets in APP. In turn, the simulations contrast emphasizes the importance of tailoring plasma deposition parameters for different liquid monomers in order to obtain a uniform coating.

Liquid precursors	Vapor species density $\times 10^{-4}$ (g cm^{-3})	Electron density $\times 10^{10}$ (cm^{-3})
HMDSO ($C_6H_8OSi_2$)	1.42	6.82
n-Hexane (C_6H_{14})	1.21	6.26
TEOS ($C_8H_{20}O_4Si$)	0.63	6.02
Water (H_2O)	0.25	5.16

Table 6.
Maximum values of vapor species density and electron density.

4.2.2 Impact of precursor flow rates

In assessing the impact of different flow rates to achieve the desired features of the surface coating deposition, the initial size distribution of the HMDSO droplets is pivotal. In discharge plasma, droplets are introduced with similar size distributions in the domain ($4\ \mu m \le r_d \le 18\ \mu m$) at the start. Using the laser diffraction particle size analysis technique as employed by atmospheric pressure plasma jet deposition, similar initial size distributions of droplets in the domain ($4\ \mu m \le r_d \le 18\ \mu m$) are introduced in the discharge plasma. According to numerical simulations [56], there is only a small amount of mutual interaction between the droplets at low precursor flow rates ($<100\ \mu l\ min^{-1}$). However, these interactions are important for altering the structure of discharge plasma at higher precursor flow rates. To demonstrate the significance of collective collision interactions between droplets and explore the effects on the atmospheric pressure plasma distribution, the fluid-droplet model is investigated at 200 and 500 $\mu l\ min^{-1}$ under identical numerical framework.

The droplet size distributions at 0 ms are represented with black bars, whereas the density bars at 200 and 500 $\mu l\ min^{-1}$ are depicted with blue and red bars, respectively, as shown in **Figure 9(a)**. Vaporization of droplets is responsible for the shrinkage of their size domains, along with the events of mutual interactions involving grazing and coalescence, respectively, at 200 $\mu l\ min^{-1}$ at 1 ms. Based on the comparison of size distributions of droplets with blue and red bars, it is apparent that the collision events are substantially lower at 200 than 500 $\mu l\ min^{-1}$. The droplet radii increment more rapidly as the coalescence of droplets continues to amplify at 500 $\mu l\ min^{-1}$, and it clear from the size domain ($2.5\ \mu m \le r_d \le 40\ \mu m$) of droplet radii relative to the initial size of droplets as shown in **Figure 9(a)** with black bars. Droplets scatter and settle during descent in the plasma chamber as they pass through these collision events gradually diminishing with time.

In the previous research work [56], they observed that the droplets with larger radii are pulled toward the substrate surface by gravity and electric potential that attract the droplets toward substrate surface. The evaporation of droplets continues to squeeze the radii of droplets in the plasma chamber as the droplets attain the steady state during downward drag at 500 $\mu l\ min^{-1}$. In spite of mutual interactions making a significant contribution to droplet coalescence, the size domain of droplets has

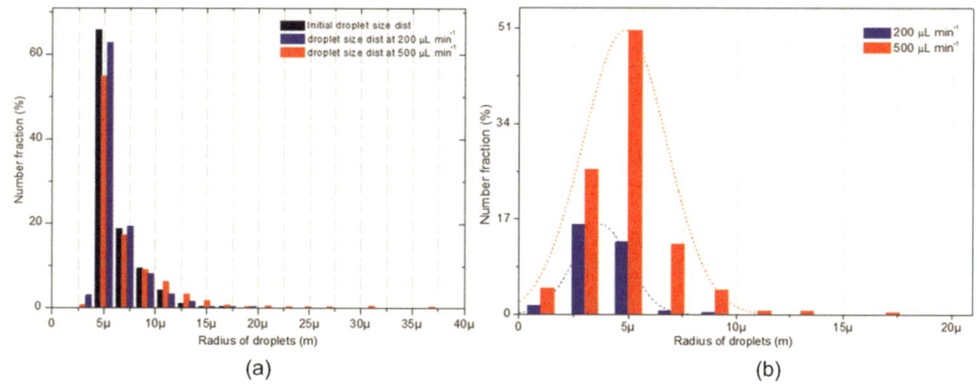

Figure 9.
(a) Black bars corresponds to 0 ms and blue as well as red at 1 ms. (b) Distribution of droplets radius at 12 ms using hexamethyldisiloxane using 200 and 500 $\mu l\ min^{-1}$ precursor flow rates, 5 $l\ min^{-1}$ gas flow rate, and 20 kHz as frequency (f) and $V_o = -13.5\ kV$.

become smaller, and it occurs within the range (100 nm $\leq r_d \leq$ 18 μm) at 12 ms in both cases as marked in **Figure 9(b)**. The smaller domain of droplet radii is caused by the ejection of higher radii of droplets from the plasma chamber indicating with blue bars at 200 μl min^{-1} with a smaller population as well as size domain in contrast to red bars at 500 μl min^{-1} as shown in **Figure 9(b)**. A large part of the droplets cross the plasma channel without evaporating completely and the remaining droplets found a decent opportunity to evaporate strongly in APP during downward drag. In the plasma chamber, droplets are found in the nanometer range prior to phase transition to vapors as evidenced by the multiple bar distributions.

4.2.3 Spatial distributions of electrons and N_2 ions

It has been observed that many nonthermal APP characteristics are associated with the presence of trace quantity of N_2 impurities in the helium gas because these impurities trigger a reduction of the ignition potential by the action of Penning ionization [29, 45]. To determine whether Penning ionization dominates in forming charge carriers along the pulse of droplets, the spatial profiles of charge carriers can provide a better understanding at the peak value of discharge current density at 12 ms to highlight its impact. As compared to the plasma chamber, He^+ and He_2^+ ions densities are greater near the momentary cathode electrode during the breakdown phase in contrast to N_2^+ ions around the pulse of droplets indicating coercive effect of Penning ionization.

Figure 10(a, a′)showed the 3D spatial structure of electrons and N_2^+ ion species at the maximum value of discharge current density along the radial and axial axes at 200 and 500 μl min^{-1}, respectively. This happens by the evaporation of droplets around the pulse. As depicted in **Figure 10**, the mass flow of liquid precursor in discharge plasma at 500 μl min^{-1} is greater than 200 μl min^{-1} because of broader effective area of evaporation at 500 μl min^{-1}. The density of electrons surrounding the droplets is slightly greater than the density of N_2^+ ions, implying that electrons generate through multiple channels, such as direct ionization, Penning ionization, and stepwise ionization. Unlike N_2^+ ions, He and He_2^+ ions densities are smaller around the pulse of droplets in the plasma chamber which highlights an impact of Penning ionization. A uniform layer is formed on the substrate surface if the evaporation effect is spread evenly along the radial axis. There is a higher probability at 500 μl min^{-1} because of the formation of uniform sheet of electrons along the radial axis as opposed to 200 μl min^{-1}. According to the spatial distribution of charge carriers, a faster precursor flow rate results in greater density values at 500 μl min^{-1}, underlining why higher precursor flow rates result in higher density values. This discussion clarifies that different precursor flow rates in APP influence the kinetics of charge carriers.

4.2.4 Impact of gas flow rates

A classification of the behavior of complex interactions between liquid droplets and APP based on similar initial size distributions of droplets in the domain is made by considering different gas flow rates with the domain's initial size distributions (5 μm $\leq r_d \leq$ 24 μm). At the inlet boundary of the PlasmaStream system, the gas flow is laminar, but the time it takes before droplets acquire settling velocity differs with gas flow rates during the downward fall. In APP, the properties of droplets and charge carriers are dynamically modified at different gas flow rates, such as 5, 10, and 20 l min^{-1} that are exhibited with the spatial distribution of vapor species and electron densities as shown in **Figure 11**. Evaporation length is longer at 20 l min^{-1} than at 5

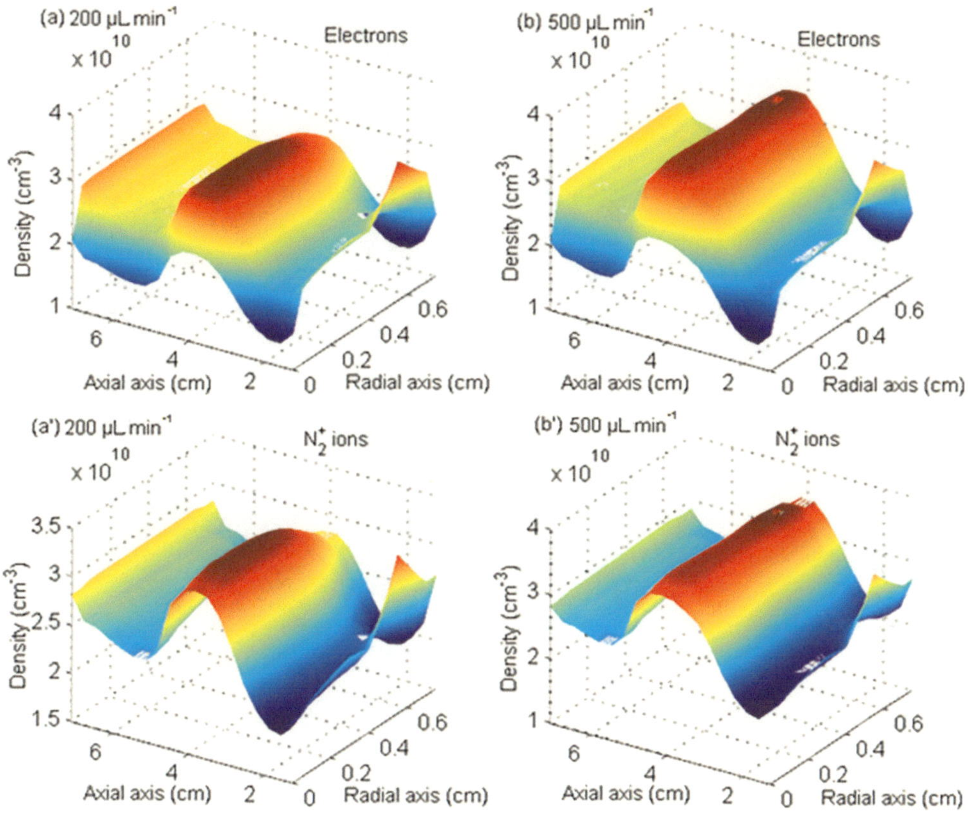

Figure 10.
3D profiles of electron density (a, b) and N_2^+ ion density (a′, b′) for HMDSO precursor droplets at flow rates of 200 and 500 μl min^{-1}, gas flow rate = 5 l min^{-1}, f = 20 kHz, and V_o = −13.5 kV.

and 10 l min^{-1} gas flow rates. With greater gas flow rates, this amplification develops because evaporation and convection of gas mixtures occur in the plasma chamber. It takes around 2 ms for an entire pulse of droplets to evaporate at a gas flow rate of 20 l min^{-1}, which is smaller than the gas flow rates at 5 and 10 l min^{-1}. The spatial profiles of the densities of the vapor species in **Figure 11(a–c)** demonstrate how the rate of cooling shifts quickly toward the exit axis when the gas flow rate increases.

In order to determine the effect of different gas flow rates on electron distribution during droplet-plasma interactions, **Figure 11(a′–c′)** exhibits the distribution of electrons during the droplet-plasma interaction. In plasma channel, the evaporation rate is greater locally when droplets travel at a low gas flow rate, such as 5 l min^{-1}. When the surrounding region of droplets is saturated due to evaporation, the remaining droplets in the pulse evaporate at a lower rate due to the impact of this saturated environment. As time passes in discharge plasma, the effect of evaporation spreads across it, increasing the rate of evaporation again. This process is relatively slow at 5 l min^{-1} as compared to 10 l min^{-1} and 20 l min^{-1}. Since gas has a short residence time, its convection is agile at higher gas flow rates, causing a sharp evaporation of droplets. This means that there are smaller chances of saturation due to higher gas flow rates, such as 20 l min^{-1} which can be confirmed from the increment in the density of electrons from ~3.50 × 10^{10} cm^{-3} to 4.50 × 10^{10} cm^{-3}, when the gas flow rate varies from 5 to 20 l min^{-1}. Due to the intense gravitational pull, evaporation of large radius drops in the

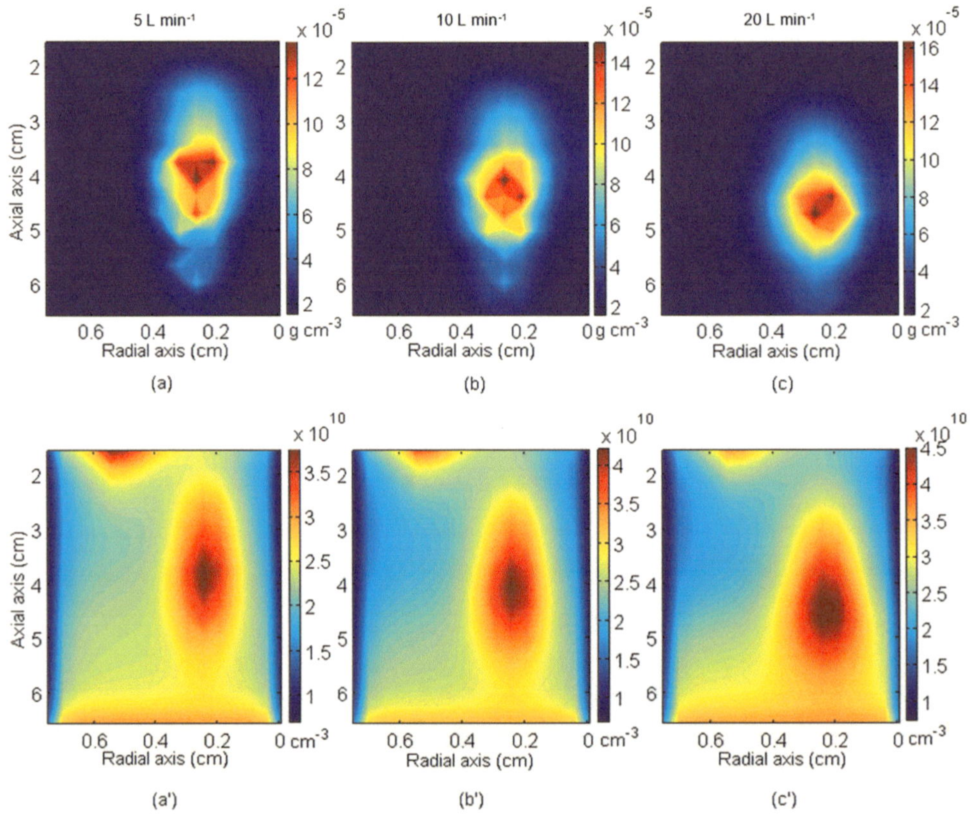

Figure 11.
Distributions of (a, b, c) vapor species density and (a′, b′, c′) electron density at precursor (HMDSO) flow rate at 100 μl min^{-1} using gas flow rates = 5 l min^{-1}, 10 l min^{-1}, and 20 l min^{-1}, f = 20 kHz, and V_o = −13.5 kV.

plasma chamber occurs within a short period of time, which causes their journey to be accelerated. However, the small duration of evaporation at 20 l min^{-1} indicates the fast conversion of liquid material into vapor phase. According to these simulation outcomes of vapor species and electron densities, it is clearly observed that the characteristics of surface deposition can be readjusted by the control of the gas flow rates.

4.2.5 Effect of gas flow rates on gas temperature and electron mean energy

By applying similar conditions, the spatial patterns of gas temperature and electron mean energy are investigated at distinctive gas flow rates. While falling, the droplets reach a stationary state, indicating that the initial velocity is most important for mutual interactions, rather than after settling in the plasma chamber. In the process of transport of two-phase flow, the local mass flux ratio ($\rho_d u_d/\rho_{gas} u_{gas}$) alters until the droplets are all converted into vapor phase. This emerges as the reduction of gas temperature in the vicinity of droplets pulse by the evaporation process. **Figure 12(a–c)** reveals that as the rate of gas flow amplifies, the perturbation caused by the droplets in the mixture enhances. A change in the rate of gas flow can modify the interaction between the droplets and discharge plasma; however, the presence of impurities in the gas in two-phase flow can also play a significant role at atmospheric pressure. Impurities such as nitrogen (N_2) and oxygen (O_2) can

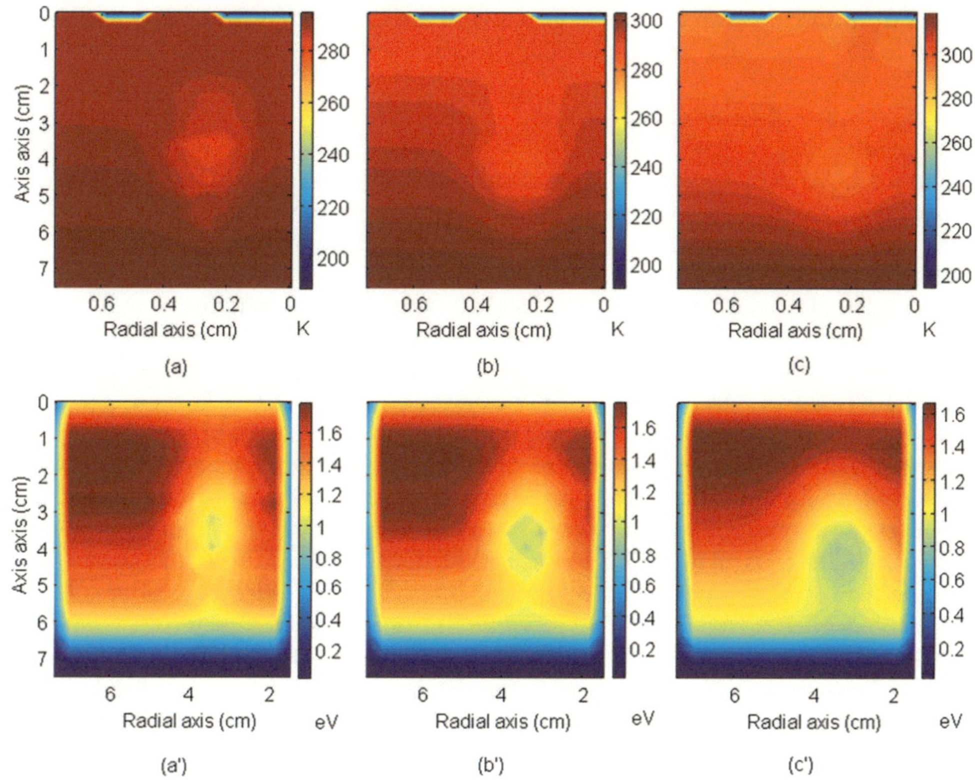

Figure 12.
Contour distributions of (a, b, c) temperature of gas and (a′, b′, c′) electron mean energy using HMDSO droplets at different gas flow rates = 5, 10, and 20 l min^{-1}, precursor flow rate = 100 μl min^{-1}, f = 20 kHz, and $V_o = -13.5$ *kV in He-N_2.*

enhance the rate of ionization because they can lead to an increase in the vaporization rate of droplets. The gas mixture ingests heat energy from the discharge plasma, and thereafter, it results as a substantial improvement in the evaporation of droplets in APP. As a result, the temperature of the gas falls along with the pulses of droplets from 5 to 20 l min^{-1} and the duration of evaporation is increased with an increment in the gas flow rate.

The temperature of electrons is calculated by the numerical solution of electron energy density equation considering the elastic and inelastic collision energy losses as well as joule heating as employed in [43, 45]. The maximum values of discharge current density during alternate cycles of discharge current pulses are observed near the thin cathode electrode and grounded substrate when the electron mean energy reaches the peak, although the prime focus lies in this case is to gain an understanding for the dynamic updates in the structure of electron mean energy along the pulse of droplets. The regions adjacent to the cathode and grounded substrate are excluded in this scenario in order to investigate the effects of evaporation on the electrons in the bulk of APP. The mean energy of electrons is reduced near the evaporating pulse of HMDSO droplets, as demonstrated by the spatial distributions in **Figure 12(a′–c′)**. Because, the spatial spread of electron mean energy is wider at 20 l min^{-1} than at 5 and 10 l min^{-1}, resulting a quicker cooling transfer downward in the plasma chamber at a higher gas flow rate. This demonstrates that the distribution of discharge plasma changes throughout droplet transit and evaporation. The spatial structures of gas temperature and electron mean energy show that when the gas flow rate increases from 5 to 20 l min^{-1}, the length of cooling increases in the plasma chamber.

4.2.6 Significance of He and He-N_2 gases

Figure 13(a) exhibited the average distribution of generation rates of stepwise ionization in the pure He and He-N_2 gases. The main sources of ionization are direct and stepwise mechanisms in He gas, while Penning ionization is more important in the He-N_2 gas mixture than other generation rates of ionization along the pulse of droplets. The stepwise ionization process is illustrated as a dotted line in this case and is responsible for ionization in He gas due to the abundance of a high density of metastables. The numeric value of the rate of stepwise ionization is $\sim 2.0 \times 10^{14}$ cm^{-3} s^{-1} in pure helium gas, and it curtails to $\sim 1.0 \times 10^{12}$ cm^{-3} s^{-1} in He-N_2 gas mixture. **Figure 13(b)** shows the rate of net ionization in atmospheric pressure plasmas, which is the outcome of direct and Penning ionization processes. Through Penning ionization, He-N_2 gas mixture experiences a faster rate of metastable destruction. It eliminates the vast majority of metastables, increasing the rate of ionization in large parts of APP. As a result, the density of metastables is reduced to a small amount in He-N_2 gas. The net ionization rate in He-N_2 discharge plasma is imperatively greater than pure He gas, while its numeric values $\sim 2.0 \times 10^{15}$ cm^{-3} s^{-1} in He-N_2 diminishe to $\sim 1.0 \times 10^{12}$ cm^{-3} s^{-1} in pure He gas as shown in **Figure 13(a, b)**. The imbalance in the rate of net ionization is due to Penning ionization caused by the presence of trace quantity of nitrogen impurities, and this is highlighted in **Figure 13(b)** by a solid line. Droplets in the plasma chamber vaporize rapidly due to the impact of net ionization in He-N_2 gas mixture, and therefore their lifetime is reduced. Based on the above outcomes, it is clearly observed that the rate of evaporation is amplified in He-N_2 gas mixture as compared to pure helium gas, which is consistent with the previous [19] simulation modeling results.

To explore the behavior of plasma species, **Figure 14** showed the line-averaged distributions electrons, He_2^+ ions, N_2^+ ions, and metastables (He^*) species density over various cycles in the pure helium using solid black lines plus symbols, while showing the red and green lines plus symbols in He-N_2 gas mixture. These distributions are obtained by neglecting the areas near the cathode and grounded substrate to explore the performance of droplet-plasma interaction during transport in the plasma

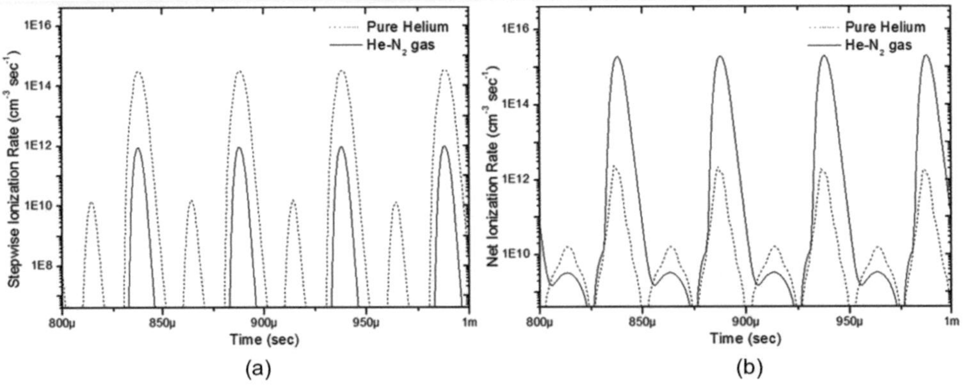

Figure 13.
Distribution of averaged generation rates of (a) stepwise ionization and (b) net ionization in the pure He and He-N_2 gases using gas flow rate = 10 l min^{-1} and HMDSO precursor flow rate = 100 μl min^{-1}, f = 20 kHz, and V_o = −13.5 kV.

chamber. The direct excitation is diagnosed as the central mechanism in the pure helium and He-N_2 gas mixtures, but the relevant contribution of metastables becomes divergent in both gas mixtures. As can be seen from the line distribution of species density in **Figure 14(a, b)**, electrons and molecular helium ions are responsible for keeping quasi-neutrality in pure helium gas. On the other hand, the presence of nitrogen impurity molecules in He-N_2 gas changes the situation, resulting in the existence of N_2^+ ions and electrons in APP, as shown in **Figure 14(a′, d′)**. This clearly shows that the Penning process is the primary ionization mechanism for supplying electrons along the pulse of droplets in He-N_2 discharge plasma. The line-averaged behavior of electrons suggests that the ionization rate is amplified by the evaporation of droplets because of the high impact of Penning ionization than direct and stepwise ionization rates. The temporal distribution of metastables shows that they have a higher density in pure helium than in He-N_2 gas mixture, as seen by the black line with hollow triangle and red line plus solid triangles in **Figure 14(c, c′)**. This occurs due to higher destruction rate of metastables through Penning ionization as compared to stepwise ionization. Therefore, the above discussion showed that the major role of interaction between the droplets and plasma depends on the chemical reactions that are happening during two-phase flow.

Under comparable operating conditions, **Figure 15** shows a clear distinction in the spatial distributions of electron density in pure helium and He-N_2 gas mixtures. At the highest value of discharge current density, the contour spatial distributions of electron density are exhibited in both gas mixtures. In the plasma chamber, the distributions of electron density are shown in **Figure 15(a, b)**, whereas the bottom row depicts the distribution of HMDSO droplets throughout the pulse as illustrated in **Figure 15(a′, b′)**, neglecting the thin cathode electrode and grounded substrate. The electron dynamics along the pulse of evaporating droplets are made clearer using this method. The electron density is significantly greater near the thin cathode electrode than along the pulse of droplets in the case of pure He, but its magnitude falls from 1.91×10^{11} to 1.22×10^{10} cm^{-3} as shown in **Figure 15(a, a′)**. The similar trend in the contraction of electron density is observed in He-N_2 gas mixture in which it alters from 3.15×10^{11} to 5.55×10^{10} cm^{-3} as displayed in **Figure 15(b, b′)**. It is clear from the shift in the magnitudes of the electron density that its value (2.595×10^{11} cm^{-3}) in He-N_2 mixture

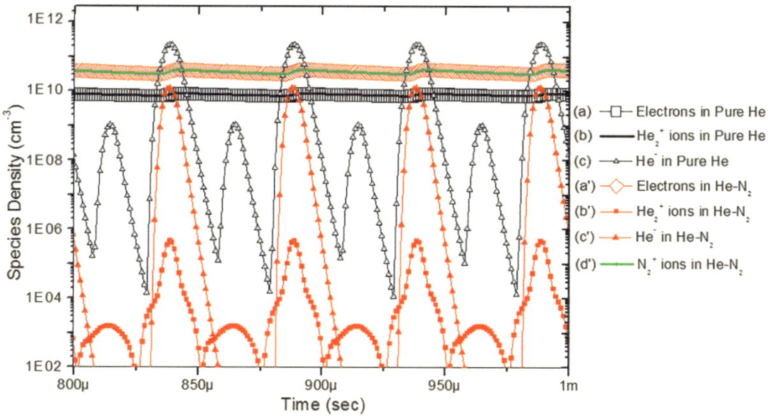

Figure 14.
Comparison of line-averaged species density (electrons, He_2^+ ions, N_2^+ ions, and He) in the pure helium and He-N_2 gas mixtures using HMDSO droplets at precursor flow rate = 100 µl min^{-1}, gas flow rate = 10 l min^{-1}, f = 20 kHz, and V_o = −13.5 kV.

is greater than the value (1.788×10^{11} cm^{-3}) in the pure helium gas. This suggests that the rapid rate of evaporation is accelerated by the significant ionization effect in the He-N_2 gas mixture, which accelerates the desolvation of droplets in the APP. As a result, this creates a feasible environment in discharge plasma to achieve the desired uniform deposition coating.

According to simulation outcomes, it has been rarely identified the decomposition of HMDSO precursor droplets at small precursor flow rates (10–100 μl min^{-1}), which can be confirmed by considering the criterion of Rayleigh limit. In APP, the magnitude of surface charge on the droplets in the mentioned size domain of radii is less than Rayleigh limit as discussed in [24]. The area of ionization around the pulse of droplets is large in He-N_2 gas, while it squeezes in pure helium gas combination, according to the spatial profiles of electron density. This means that the composition

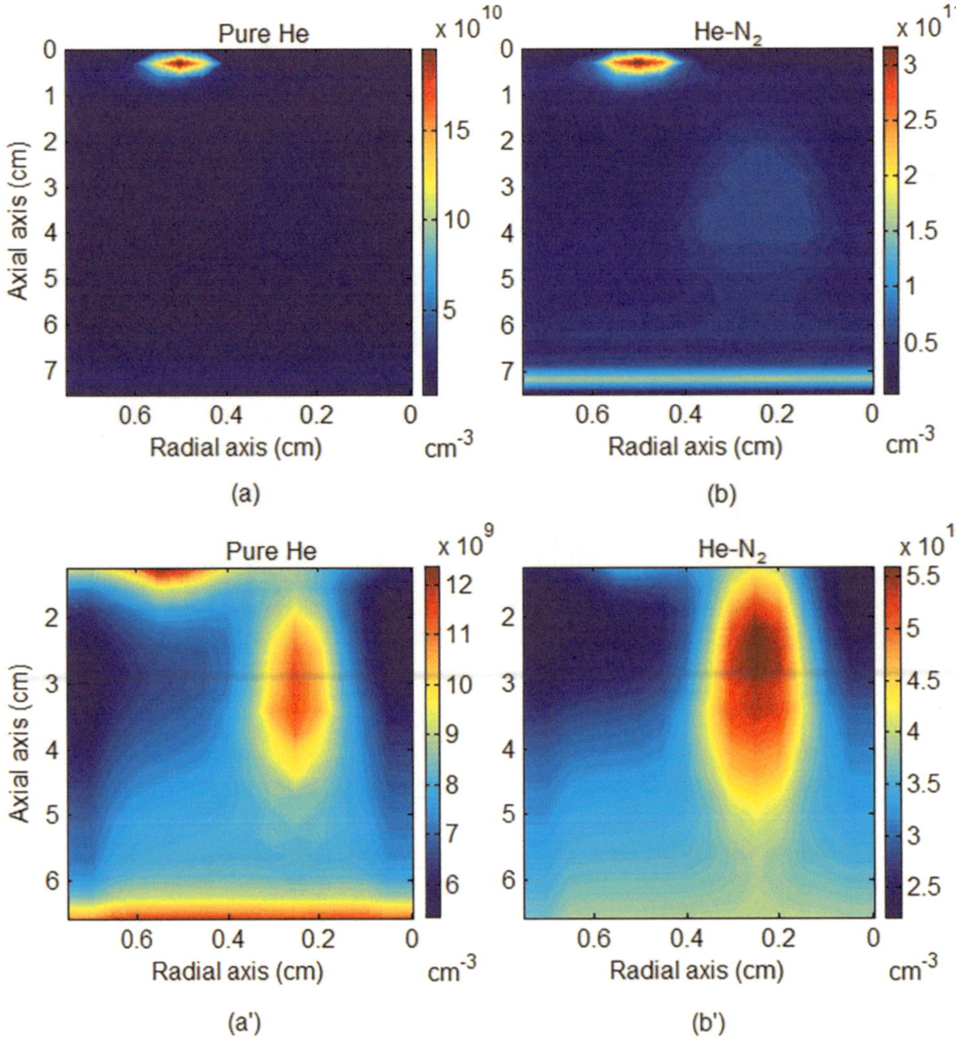

Figure 15.
Profiles of electron density in the plasma chamber (a, b), electron density around the pulse of HMDSO droplets (a′, b′) in the pure He and He-N_2 gases at gas flow rate = 10 l min^{-1}, HMDSO precursor flow rate = 100 μl min^{-1}, f = 20 kHz, and V_o = −13.5 kV.

of the operating gas mixture affects droplet evaporation. If the gas mixture contains small amount of impurities that are capable of modifying the properties of discharge plasma, and ultimately, the evaporation and mutual interactions of droplets are effected. In APP, the droplets acted as the perturbing agent at the beginning; however, they became part of the operating gas mixture after phase transformation later as vapors. The entire set of interactions is glued through the coupling source terms providing a clear description of heat transfer in two-phase flow. The characteristics of cold plasma can be manipulated by the control of precursor and gas flow rates, because the rate of evaporation inflates with an increment of precursor and gas flow rates. This kind of cold APP is identified as suitable for the medical and industrial applications [6, 9, 25, 26]. In case of incomplete evaporation of droplets, they are not considered as appropriate for the coating deposition applications due to the presence of droplets in discharge plasma. As can be seen from the above contrast, N_2 impurities are extremely helpful in enhancing ionization activities during transport of carrier of gas in the plasma chamber, which therefore enhances vaporization of droplets, resulting in a homogenous plasma environment appropriate for surface coatings.

4.2.7 Comparison of model and experimental observations

To verify the numerical simulation results, it is critical to do a comparison with experimental measurements using a separate setup of laser diffraction particle size analysis technique with the PlasmaStream atmospheric pressure jet deposition system. At location B in the PlasmaStream system in **Figure 5(b)**, a pulse of HMDSO precursor droplets is injected within the size range of $1\ \mu m \leq r_d \leq 6\ \mu m$ in APP. At small precursor flow rates ($\leq 100\ \mu l\ min^{-1}$) [56], the mutual interactions between HMDSO droplets, such as grazing and coalescence, are minimal. The fluid-droplet model developed in this chapter's simulation study had similar experimental settings as those listed in **Table 3**, and the initial injection velocity of droplets was assumed to be the same as $1.5 \times 10^3\ cm\ s^{-1}$ in this case. The radii of droplets are reduced steadily by the main contribution of evaporation on the surface during downward fall in the plasma chamber.

The initial distribution of droplets exists in the range ($\geq 1\ \mu m$) before introducing into plasma chamber, and the size radii of droplets continuously contract due to evaporation as clearly observed by the presence of droplets within nanometer range ($100\ nm \leq r_d \leq 900\ nm$) as shown in **Figure 16(a)** at 7 ms. This demonstrates that the HMDSO liquid precursor in He-N_2 discharge plasma is highly volatile. The experimental size distributions are recorded in two different experiments in the plasma chamber as shown in different colors in **Figure 16(b)**, but the size domain of droplet radii is similar in both cases. It is evident from the size domain of droplets as mentioned in dotted curly brackets in **Figure 16(a, b)** that the entire bunch of droplets lies in the domain ($500\ nm \leq r_d \leq 5\ \mu m$) at 7 ms, which exhibits a good agreement between the coupled fluid-droplet model and experimental size distributions. In the experimental measurements, the split at 0.5 μm is a result of the limiting resolution of the laser diffraction imaging lens arrangement. However, as seen in the size distribution of droplets, the fluid-droplet model also offers information regarding the occurrence of minimum feasible radii of droplets in the nanoscale range as highlighted in **Figure 16(a)**. The foregoing comparison clearly demonstrates that under similar operational constraints, the numerical model and experimental measurements are synchronized nicely. As a result, the similarity of results increases confidence that the numerical simulations utilizing the 2D-coupled fluid-droplet model can accurately describe the complicated interaction between two-phase flow.

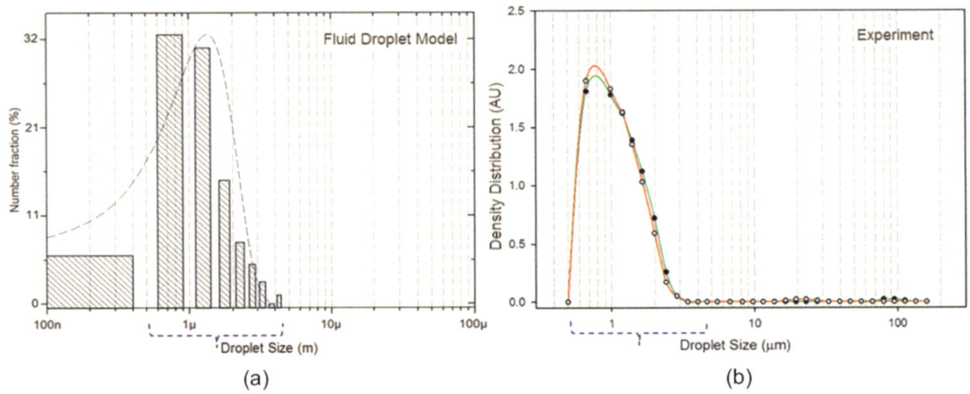

Figure 16.
Comparison of (a) fluid-droplet model and (b) experimental observations of size distribution of droplets using HMDSO precursor droplets at flow rate = 100 μl min^{-1} and gas flow rate = 5 l min^{-1}, f = 20 kHz, and $V_o = -13.5$ kV.

4.2.8 3D profiles of droplet-plasma interaction

The dynamic characteristics of evaporation utilizing TEOS and HMDSO precursors under the same limitations are described in this section using three-dimensional profiles species density in APP. The validity and legitimacy of multidimensional (2D and 3D) numerical modeling outcomes have already been described in [43] by comparing with experimental measurements. **Figure 17** shows the iso-contours of electrons (red) and vapor species density (green) at three consecutive time instants (t_1 = 0.6, t_2 = 1.6, and t_3 = 2.6 ms) during the evaporating pulse of two distinct precursor droplets (TEOS and HMDSO). In the case of both precursors, the iso-contours of vapor species density revealed that their volumes grow with time; however, in the discharge plasma, TEOS expands in volume faster due to higher evaporation of droplets than HMDSO. Because the mass density of TEOS and HMDSO droplets differs, the bunch of TEOS droplets experiences a stronger gravitational attraction than the HMDSO droplets, and eventually, it manifests itself as a disparity in their volumetric spread as clearly contrasted in the top and bottom rows of **Figure 17**. As shown by the blue dashed line, the affected volume along the pulse of droplets extends and pulls downward. On the other hand, the rate of vaporization of HMDSO droplets is greater than TEOS in He-N_2 discharge plasma because of its less surface tension and boiling point as compared to TEOS. This is supported by **Table 5** that showed greater values of vapor species density for HMDSO than TEOS, while these features are in accordance with the previous results discussed in the previous subsections using two-dimensional-coupled fluid-droplet model.

At t_1, the conducting channel is forming an intensive ionization along the pulse of both precursor droplets, as shown in **Figure 17(a, a′)**. At t_2, **Figure 17(b, b′)** exhibits how the overlapped density layers of electrons (red) and vapor species density (blue) are expanding, with a growth of vapor species continuing until the full pulse of droplets dissolves in the discharge plasma. In terms of electron kinetics, they are present throughout the plasma chamber at the highest value of negative discharge current density, as illustrated in **Figure 17** by the red iso-contours. This demonstrates a progressive increase in electron density in the plasma chamber. The numerical values listed in **Table 5** indicate that the electron

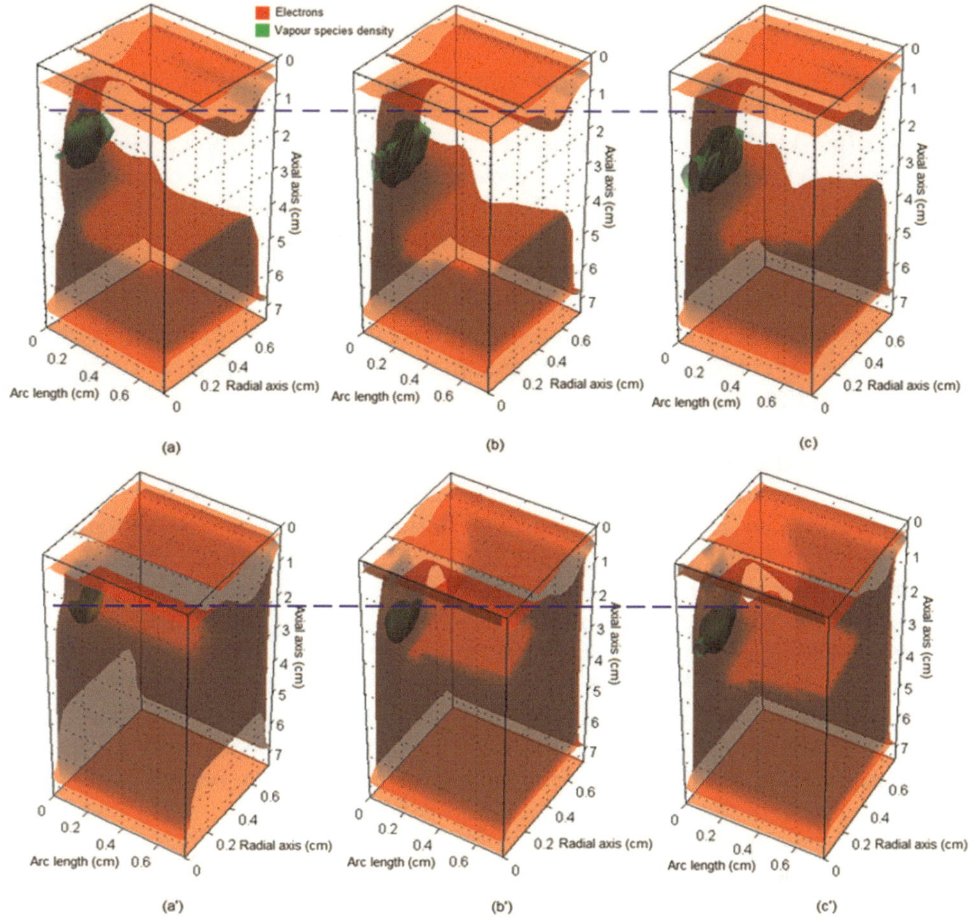

Figure 17.
Spatio-temporal iso-contours of electrons (red) and vapor species density (green) using TEOS (top row, a, b, c) and HMDSO (bottom row, a′, b′, c′) at t_1, t_2, and t_3 ms at precursor flow rate = 100 µl min^{-1}, gas flow rate = 5 l min^{-1}, f = 20 kHz, and V_o = −13.5 kV [44].

density around the pulse of HMDSO droplets is higher than TEOS. HMDSO has a smaller effect on the volume of vapor species compared with TESO due to the robust evaporation rate in APP. Additionally, the numerical simulation outcomes highlighted that the duration of evaporation for the entire pulse of TEOS droplets is almost twice as long as its counterpart for HMDSO droplets. Thus, TEOS droplets scatter more than HMDSO droplets causing the volume of vapor species to increase. It is evident that these observations differ sharply from those of the top and bottom rows as shown in **Figure 17**. Following the discussion above, it is evident that liquid droplets respond very differently to discharge plasma, depending on the type of precursor. As a result, the numerical modeling outcomes are deemed to be adequate for describing a good understanding of complex interaction of droplets with plasma in APP. These quasi-volumetric characteristics suggest that the internal structure of discharge plasma in two-phase flow is highly complicated, which may be further rationalized by considering a complex chemistry set comprising chemical reactions between vapors and discharge species.

5. Conclusion

In this chapter, we discussed how droplet-plasma interaction is initiated in the atmospheric pressure plasmas. For describing the properties of plasma and explain droplet transport and interaction with the plasma, the one-dimensional normalized fluid model is developed considering Boltzmann distribution of electrons around the droplets. Based on the simulation outcomes, it is concluded that the evaporation and charging of droplets are considered as the most dominant mechanisms within the plasma channel. After developing and testing multidimensional fluid models (2D and 3D) in two-phase flow under distinct precursor and gas flow rates, we examined the implications of complicated interaction between the droplets and plasmas. The characteristics of droplet-plasma interaction manifest that the evaporation of droplets is noticed as the prime mechanism, which has been verified by using different types of liquid precursors (HMDSO, n-hexane, TEOS, and water). A deeper observation at the mean profiles of droplet radii showed that the mutual interactions between the droplets are a major factor to modify the structure of discharge plasma in the limit of higher precursor flow rates ($\geq$200 μl min^{-1}). Additionally, the spatial and temporal distributions of droplets and APP are explored to understand the interactive behavior of various precursors, showing that evaporation is remarkably intensive in the case of HMDSO and n-hexane compared to TEOS and water. In the plasma chamber, the convection of cooling is much shifted downward at 20 l min^{-1} compared with smaller gas flow rates, justifying the significance of the evaporation process at different gas flow rates (5, 10, and 20 l min^{-1}). To establish the validity of the numerical simulation results, we compared the outcomes from the 2D fluid-droplet model with experimental observations. Finally, a 3D view of droplet-plasma interaction is presented using spatio-temporal iso-contours of electrons and vapor species density using HMDSO and TEOS to show the volumetric spread in APP.

Acknowledgements

This work is supported by Science Foundation Ireland under grant 08/SRC/I1411.

Author details

Muhammad M. Iqbal[1*] and Mark M. Turner[2*]

1 CCT College Dublin, Dublin, Ireland

2 School of Physical Sciences, Dublin City University, Dublin, Ireland

*Address all correspondence to: miqbal@cct.ie and miles.Turner@dcu.ie

References

[1] Fridman A. Plasma Chemistry. New York, ISBN: 13 978–0–521-84735-3 Hardback: Cambridge University Press; 2008

[2] Becker KH, Kogelschatz U, Schoenbach KH, Barker RJ. Non-Equilibrium Air Plasmas, Series in Plasma Physics. Bristol and Philadelphia, ISBN: 0 7503 0962 8: IOP Publishing; 2005

[3] Reece Roth J. Industrial Plasma Engineering. Vol. 2. Bristol and Philadelphia, ISBN: 0 7503 0544 4: Institute of Physics Publishing; 2001

[4] Sanchez-Estrada FS, Qiu H, Timmons RB. Molecular tailoring of surfaces via RF pulsed plasma polymerizations: biochemical and other applications. IEEE Conference Record - Abstracts. 2002 IEEE International Conference on Plasma Science (Cat. No.02CH37340). 2002. p. 254. DOI: 10.1109/PLASMA.2002.1030530

[5] Kolb JF, Mohamed A-AH, Price RO, Swanson RJ, Bowman A, Chiavarini RL, et al., Cold atmospheric pressure air plasma jet for medical applications. Applied Physics Letters. 2008;**92**: 241501

[6] Laroussi M. Nonthermal decontamination of biological media by atmospheric-pressure plasmas: review, analysis, and prospects, in IEEE Transactions on Plasma Science. Aug 2002;**30**(4):1409-1415. DOI: 10.1109/TPS.2002.804220

[7] Stoffels E, Kieft IE, Sladek RE J. Superficial treatment of mammalian cells using plasma needle. Journal of Physics Department. 2003;**36**:2908–2913. DOI: 10.1088/0022-3727/36/23/007

[8] Pierre-Luc G-L, Mwale F, Iordanova M, Demers C, Desjardins P, Wertheimer MR. Atmospheric Pressure Deposition of Micropatterned Nitrogen-Rich Plasma-Polymer Films for Tissue Engineering. Volume 2, Issue 3, Special Issue: Highlights from the Ninth International Symposium on High-Pressure, Low-Temperature Plasma Chemistry 2004 in Padua, Italy. March 31, 2005, pp. 263-270

[9] Fridman G, Peddinghaus M, Ayan H, Fridman A, Balasubramanian M, Gutsol A, et al., Blood coagulation and living tissue sterilization by floating electrode dielectric barrier discharge in air. Plasma Chemistry and Plasma Processing. 2006; **26**:425-442. DOI: 10.1007/s11090-006-9024-4

[10] Yu Babaeva N, Kushner MJ, Intracellular electric fields produced by dielectric barrier discharge treatment of skin. Journal of Physics Department. Applid Physics. 2010;**43**:185206

[11] Massines F, Rabehi A, Decomps P, Gadri RB, Segur P, Mayoux C. Experimental and theoretical study of a glow discharge at atmospheric pressure controlled by dielectric barrier. Journal of Applied Physics. 1998;**83**:2950. DOI: 10.1063/1.367051

[12] Wang Q, Doll F, Donnelly VM, Economou DJ, Sadeghi N, Franz GF. Experimental and theoretical study of the effect of gas flow on gas temperature in an atmospheric pressure microplasma. Journal of Physics Department. Applied Physics. 2007;**40**:4202

[13] Babayan SE, Beach H, Selwyn GS, Fe S, Hicks RF. Patent No: US 6,194,036 B1, Feb. 27, 2001, Published in Official Gazette of the United States Patent and

Trademark Office. Patents February 27, 2001;**1243**(4). Published weekly by authority of congress. Available from: https://books.google.ie/books?id=xg4ETKL3IzYC&printsec=frontcover#v=onepage&q&f=false; https://patentimages.storage.googleapis.com/e4/6a/d4/d8279b5f5dcb8b/US6194036.pdf

[14] Anthony P, Herbert F, O'Neill L, Jaroszynska-Wolinska J, Stallard C, Ramamoorthy A, et al. A Comparison between Gas and Atomized Liquid Precursor States in the Deposition of Functional Coatings by Pin Corona Plasma. 22 Mar 2011;**8**(3):230-238. DOI: 10.1002/ppap.201000119

[15] Amsden AA, O'Rourke PJ, Butler TD. KIVA-II: A computer program for chemically reactive flows with sprays. United States: Np. 1989. DOI: 10.2172/6228444. Available from: https://www.osti.gov/biblio/6228444-kiva-ii-computer-program-chemically-reactive-flows-sprays

[16] Fauchais P, Vardelle A, Vardelle M. Modelling of plasma spraying of ceramic coatings at atmospheric pressure. Ceramics International. 1991:**17**(6):367-379

[17] O'Rourke PJ. Collective Drop Effects in Vaporizing Liquid Sprays. Ph.D Thesis 1532-T, Princeton University, Department of Mechanical and Aerospace Engineering. 1981

[18] Dukowicz JK. A particle-fluid numerical model for liquid sprays. Journal of Computational Physics. Apr 1980;**35**(2):229-253. DOI: 10.1016/0021-9991(80)90087-X

[19] Stallard CP, Iqbal MM, Turner MM, Dowling DP. Investigation of the Formation Mechanism of Aligned Nano-Structured Siloxane Coatings Deposited Using an Atmospheric Plasma Jet, Plasma Processes and Polys. Oct 2013;**10**(10):888-903. DOI: 10.1002/ppap.201300056

[20] Yoshida T, Akashi K. Particle heating in a radio-frequency plasma torch. Journal of Applied Physics. 1977;**48**(6):2252. DOI: 10.1063/1.324036

[21] Shan Y, Mostaghimi J. Numerical simulation of aerosol droplets desolvation in a radio frequency inductively coupled plasma, Spectrochimica Acta Part B: Atomic Spectroscopy. 21 November 2003;**58**(11):1959-1977. DOI: 10.1016/j.sab.2003.09.003

[22] Clayton TC, Schwarzkopf JD, Sommerfeld M, Tsuji Y. Multiphase Flows with Droplets and Particles. Second Edition, ISBN: 978-1-4398-4050-4. Boca Raton: CRC Press, Taylor & Francis Group, NW; 2011

[23] Leiby CC, Oskam HJ. Volume Forces in Plasmas. The Physics of Fluids. 1967;**10**:1992. DOI: 10.1063/1.1762397

[24] Iqbal MM, Turner MM, Liquid Droplets in Atmospheric Pressure Plasmas. Proceeding in 63rd Annual GEC and 7th ICRP. Paris: France. Oct 2010;**55**:7

[25] Thomas M, Mittal KL. Atmospheric Pressure Plasma Treatment of Polymers: Relevance to Adhesion. Hoboken, New Jersey: John Wiley & Sons, Inc. 2013. ISBN 978-1-118-59621-0

[26] Nelea V, Li Luo CN, Demers J, Antoniou A, Petit S, Lerouge MRW, et al. Selective inhibition of type X collagen expression in human mesenchymal stem cell differentiation on polymer substrates surface-modified by glow discharge plasma, Volume 75A, Issue 1, Pages 216-223, 1 October 2005. DOI: 10.1002/jbm.a.30402

[27] Pasandideh-Fard M, Bhola R, Chandra S, Mostaghimi J. Deposition of tin droplets on a steel plate: simulations and experiments. International Journal of Heat and Mass Transfer. 1998;**41**(19): 2929-2945

[28] Al-Bataineh SA, Szili EJ, Gruner PJ, Priest C, Griesser HJ, Voelcker NH, et al. Fabrication and operation of a microcavity plasma array device for microscale surface modification. Plasma Processes and Polymers. 2012;**9**(7):638-646

[29] Zhang P, Kortshagen U. Two-dimensional numerical study of atmospheric pressure glows in helium with impurities. Journal of Physics Department Applied Physics. 2006;**39**:153

[30] Este G, Westwood WD. A quasi-direct-current sputtering technique for the deposition of dielectrics at enhanced rates. Journal of Vacuum Science & Technology. 1988;**A6**:1845. DOI: 10.1116/1.575266

[31] Birmingham JG, Moore RR. Reactive Bed Plasma Air Purification. US Patent no. 4954 320, 1990

[32] Gerenser LJ, Grace JM, Apai G, Thompon PM. Surface chemistry of nitrogen plasma-treated poly(ethylene-2,6-naphthalate): XPS, HREELS and static SIMS analysis, Surface and Interface Analysis. Volume 29, Issue 1, Pages 12-22, January 2000. DOI: 10.1002/(SICI)1096-9918(200001)29:1 <12::AID-SIA687>3.0.CO;2-7

[33] Harlow FH, Amsden AA. A numerical fluid dynamics calculation method for all flow speeds. Journal of Computational Physics. Volume 8, Issue 2, Pages 197-213, October 1971. DOI: 10.1016/0021-9991(71)90002-7

[34] Han Z, Fan L, Reitz RD. Multidimensional Modeling of Spray Atomization and Air-Fuel Mixing in a Direct-Injection Spark-Ignition Engine. SAE Paper 970884, SAE Transactions, Journal of Engines. 01 Jan 1997;**106**(3): 1423-1441

[35] Olesik JW, Smith LJ, Williamsen EJ. Signal fluctuations due to individual droplets in inductively coupled plasma atomic emission spectrometry, Analytical Chemistry. 15 Sep 1989;**61**(18):2002-2008. DOI: 10.1021/ac00193a003

[36] Hobbs SE, Olesik JW. The effect of desolvating droplets and vaporizing particles on ionization and excitation in Ar inductively coupled plasmas. Spectrochimica Acta Part B: Atomic Spectroscopy. May-June 1993;**48**(6-7): 817-833. DOI: 10.1016/0584-8547(93) 80087-B

[37] Meyyappan M. Computational Modelling in Semiconductor Processing, Pages: 363. Published by Artech House. 1994. ISBN 10: 0890067074; ISBN 13: 9780890067079. Available from: https://uk.artechhouse.com/Computational-Modeling-in-Semiconductor-Processing-P95.aspx

[38] Arfken G. Mathematical Methods for Physicists. 3rd Edition. January 1, 1985, Copyright: © Academic Press 1985, eBook ISBN: 9781483277820. Available from: https://www.elsevier.com/books/mathematical-methods-for-physicists/arfken/978-0-12-059820-5

[39] Iqbal MM, Turner MM, Oral presentation. Dynamic Interaction Between Liquid Droplets and Atmospheric Pressure Discharge, 37th EPS Conference, 21–25 June, 2010, Vol. 34A, ISBN 2-914771-62-2

[40] Coppins M. Electrostatic Breakup in a Misty Plasma. Physical Review Letters. 104, 065003 - Published. 11 February 2010. DOI: 10.1103/PhysRevLett.104. 065003

[41] Rayleigh L. On The Instability Of Jets. Proceedings of the London Mathematical Society, Volume s1-10, Issue 1, Pages 4-13, November 1878. DOI: 10.1112/plms/s1-10.1.4

[42] Yue Y, Sun J, Gunter KL, Michalek DJ, Sutherland JW. Character and Behavior of Mist Generated by Application of Cutting Fluid to a Rotating Cylindrical Workpiece, Part 1: Model Development. Journal of Manufacturing Science Engineering. August 2004;**126**(3):417-425. DOI: 10.1115/1.1765150

[43] Iqbal MM, Stallard CP, Dowling DP, Turner MM. Two-Dimensional Integrated Model for Interaction of Liquid Droplets with Atmospheric Pressure Plasma. Plasma Processes and Polymers. Volume12, Issue 11, Pages 1256-1270, November 2015. DOI: 10.1002/ppap.201500003

[44] Iqbal MM, Turner MM. Investigations of Droplet-Plasma Interaction using Multi-Dimensional Coupled Model, Contributions to Plasma Physics. Volume 55, Issue 9, Pages 627-642, October 2015. DOI: 10.1002/ctpp.201500048

[45] Iqbal MM, Turner MM. Spatial Uniformity of Atmospheric Pressure Discharges: A Simulation Study. Contributions to Plasma Physics. Volume 54, Issue 9, Pages 756-771, October 2014. DOI: 10.1002/ctpp.201400002

[46] Hagelaar GJM. Modeling of microdischarges for display technology. [Phd Thesis 1 (Research TU/e / Graduation TU/e), Applied Physics]. Eindhoven: Technische Universiteit, 2000. DOI: 10.6100/IR538700

[47] Peaceman DW, Rachford HH Jr. The Numerical Solution of Parabolic and Elliptic Differential Equations. Journal of the Society for Industrial and Applied Mathematics. March 1955;**3**(1). DOI: 10.1137/0103003

[48] Adler SL. Over-relaxation method for the Monte Carlo evaluation of the partition function for multiquadratic actions. Physical Review D. Volume 23, Issue 12, Pages 2901, 15 June 1981. DOI: 10.1103/PhysRevD.23.2901

[49] Lymberopoulos DP, Economou DJ. Modeling and simulation of glow discharge plasma reactors. Journal of Vacuum Science & Technology. 1994**A12**(4):1229-1236

[50] Yuan X, Raja LL. Computational Study of Capacitively Coupled High-Pressure Glow Discharges in Helium. IEEE Transactions On Plasma Science. Aug 2003;**31**(4):495-503

[51] WolframAlpha computational intelligence, ©2022 Wolfram Alpha LLC, Wolfram for Education. Available from: https://www.wolframalpha.com

[52] Hagelaar GJM, Kroesen GMW, Van Slooten U, Schreuders H. Modeling of the microdischarges in plasma addressed liquid crystal displays. Journal of Applied Physics, Volume 88, Issue 5, pp 2252, 20 March 2000. DOI: 10.1063/1.1287529

[53] Alves LL, Gousset G, Ferreira C. Self-contained solution to the spatially inhomogeneous electron Boltzmann equation in a cylindrical plasma positive column. Physical Review E. Volume 55, Issue 1, pp 890, 1 January 1997. DOI: 10.1103/PhysRevE.55.890

[54] Hagelaar G, Pitchford L. Solving the Boltzmann equation to obtain electron transport coefficients and rate coefficients for fluid models, Plasma Sources Science and Technology. Volume 14, Number 4, pp. 722-733,

2005. Available from: http://stacks.iop.org/PSST/14/722

[55] Petrucci RH, Geoffrey Herring F, Madura JD, Bissonnette C. General chemistry principles & modern applications. Upper Saddle River Pearson, 2007. ISBN: 0135334985

[56] Iqbal MM, Turner MM, Behavior of HMDSO Liquid Precursor Droplets in APD Plasma and Gas Mixture, Proceedings in 31st ICPIG, July 14-19, 2013, Granada, Spain

Section 4

Applications of Droplet Behaviors

Chapter 5

Bioinspired Smart Surfaces and Droplet Dynamics-A Brief Review

Raza Gulfam

Abstract

Mimicking the topographic structures and designs of living surfaces (e.g., lotus leaf, pitcher plant and beetle) onto the non-living surfaces (e.g., metallic plates, glass wafers, wood and fabrics) is known as bioinspiration. Consequently, the pristine topography of the non-living surfaces is robustly modified, known as bioinspired smart surfaces, providing novel surface regimes, i.e., wetting regimes and droplet dynamic regimes. Herein, factors affecting the droplet dynamics and its applications in bioinspired smart surfaces are presented. The droplet dynamics is a complicated phenomenon being affected by the various factors, encompassing the surface roughness, axial structural interspacing (ASI), structural apex layer (SAL), surface positioning, structural alignment, liquid droplet-surface interaction (LD-SI), and various stimuli, etc. Further, the droplet dynamics can be seen many applications, such as droplet manipulation, self-cleaning effect, design of controllable chemical reactors and electric circuits, water harvesting and condensation heat transfer, and oil/water separation, amongst others. The chapter has been mainly divided in three sections enclosed between the introduction and conclusion, comprehensively elaborating the classification of surface regimes, factors affecting the droplet dynamics and the applications at lab and industrial scales. In all, the contents are expected to serve as the guideline to accelerate advancement in the surface science.

Keywords: biomimetic, Superhydrophobic, slippery surfaces, wetting, droplets

1. Introduction

Surface science has revolutionized the modern technology based on bioinspired alternatives and novel applications, attempting to control and resolve academic, industrial and societal challenges across the globe. It is important herein to explicitly define the surface so as to remove the literature anomaly because it is elusive whether to consider the top-most region as a surface or the whole of it. On the one hand, the entity under study should be called the substrate, and its top-most region should be named the surface. On the other hand, if the entity under study is named as surface (as we see in the literature), its top-most region should better be termed as the topography, meaning that the region lying under the topography is the surface. Therefore, we are also likely to adopt and utilize the latter terminologies henceforward, i.e., the topography is the part of a surface where the water droplet interacts. Altogether, surface science is based on the fundamental surface models (Young's model, Wenzel model and Cassie-Baxter model) that are employed to understand the wettability/wetting, droplet dynamics and the involved surface engineering processes [1–3].

Surfaces can be categorized as living and non-living surfaces.

Living surfaces include birds, insects, plants and animals. Upon careful observation, it has been found that the living surfaces behave very uniquely and differently with the water droplets. For example, the rainwater cannot wet the wings of the butterfly even during rain. Water striders can move on the water surface with great ease. Water can flow on the rice leaves very fast. Water droplets can dance on the lotus leaf and finally rolls off by taking the dust with them. Water droplets can stay on the rose petals even if it is vertically standing. Fish can travel through water very efficiently, and so on. When the microscopic examination of various living surfaces is carried out, the topographic morphologies have been found different in each case, consisting of micro/nano-structures of a wide variety of geometries, arrangements and interspacing. In addition, a particular kind of layer surrounding the micro/nano-scaffolds has been discovered, which is named the epicuticular wax [4]. Therefore, a constructive conclusion can be drawn that the topography of various surfaces is the main driver that affects the water droplets. In other words, living surfaces provide different levels of adhesions and slipperiness based on the epicuticular wax.

Non-living surfaces encompass metals, wood, polymer, glass, fabric and paper, etc. Until recently, rigorous efforts have been underway, aiming to modify the topographies of the non-living surfaces with help of bioinspiration. The bioinspiration is a broad technological and scientific concept in which the micro/nano-structures of living surfaces are studied, and then similar structures are created onto the topographies of non-living surfaces, which are termed the bioinspired smart surfaces. With the bioinspired conversion of wettability (i.e., bioinspired mimicking) as depicted in **Figure 1(a–c)**, the newly-born topographic structures include extruding topographic structures (ETS) (**Figure 1b**) and intruding topographic structures (ITS) (**Figure 1c**). The major influencers for bioinspired wettability are the axial structural interspacing (ASI) and the structural apex layer (SAL). Based on the ETS/ITS and ASI, the newly-born area on the topography is called the projected area over which the contact angle is known as the apparent contact angle θ_a (°). By dividing θ_a (°) by the equilibrium contact angle θ_e (°) of the pristine surface, the roughness R can be obtained. The SAL can be defined as the sites existing on the top of ETS/ITS responsible for introducing the same properties as the epicuticular wax does in the living surfaces. Therefore, SAL can be obtained by coating certain materials having either high surface energy (providing hydrophilic or superhydrophilic characteristics such as silicon dioxide [5]) or low surface energy (providing hydrophobic or superhydrophobic characteristics such as silanes [6, 7]). An experimental case study of bioinspired smart surface, known as the slippery liquid-infused porous surface (SLIPSs), can be seen in **Figure 1(d–f)**. SLIPSs are prepared by getting inspiration from the pitcher plants.

To create ITS/ETS, there are many traditional and advanced surface engineering methods [10–12], such as chemical oxidation, chemical etching, reactive ion etching, grafting, dipping, spinning, photo-lithography, electron beam-lithography, electrodeposition, imprinting, templating, hot embossing, plasma treatment, vaporization, selective tunneling, anodizing, laser ablation, etc. Thus, by selecting the most suitable method or combination of several methods, the ETS and ITS having a wide variety of geometries can be created, for example: plate-like, wire-shaped, whisker, cone-like, square-arrays, fibre-like, vertical pillars, spikes, honeycomb-shaped, grooves, regular deep-pores, holes, channels, trenches, voids, etc., and the common feature of

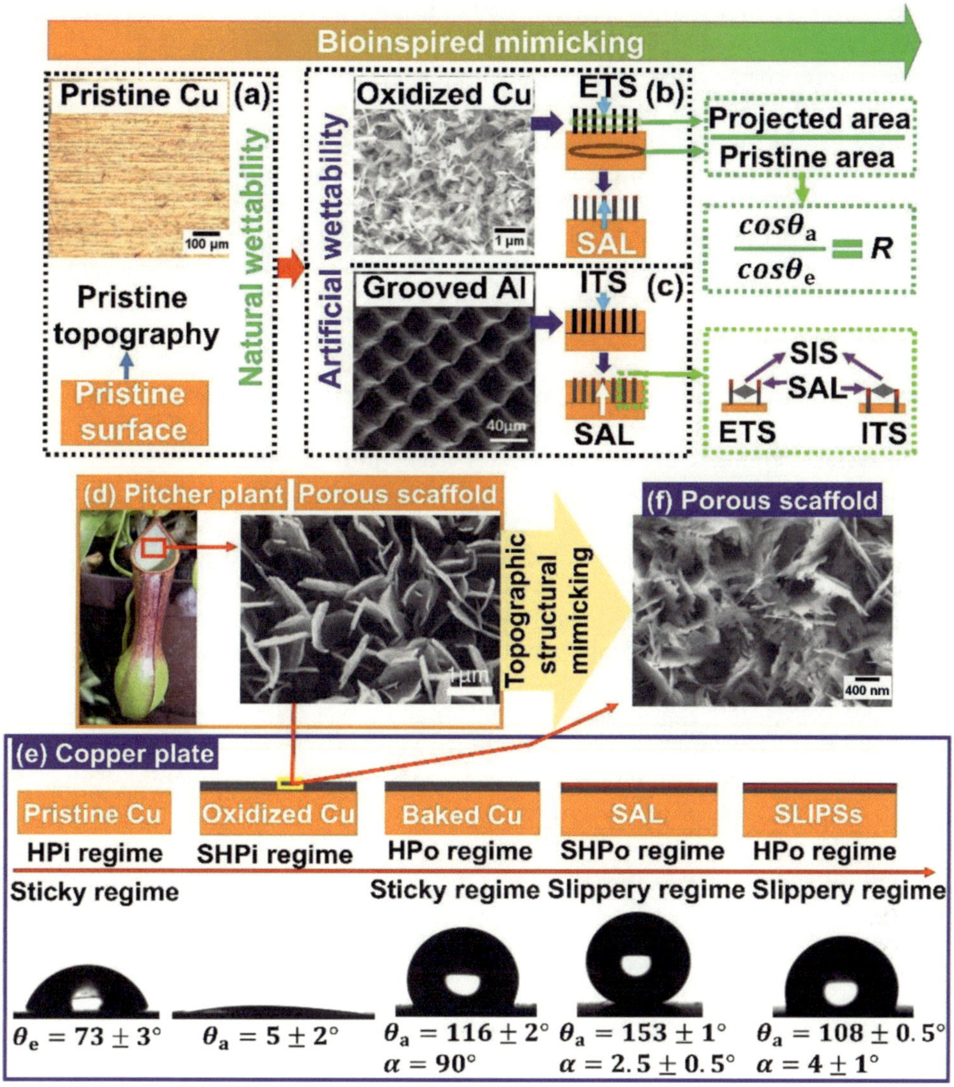

Figure 1.
Bioinspired mimicking depicting (a) pristine copper surface and topography, growth of (b) ETS and (c) ITS [8]. An experimental case study demonstrating the (d) porous scaffold of pitcher plant [9], and (e) fabrication of bioinspired SLIPSs consisting of (f) porous scaffold alike pitcher plant. [Note: Microscopic images in (a), (b) and (f), as well as the contact angles and their images, belong to the author(s)].

all geometries is their size which tends to exist at micro, nano or micro-nano mixed (hierarchical) scale [12]. Hence, droplet wetting and droplet dynamic regimes of the pristine non-living surfaces can be completely tailored. Droplet dynamics of the bioinspired smart surfaces greatly rely on the SAL and ASI of the ITS/ETS.

In brief, the bioinspired smart surfaces are more efficient and capable of controlling industrial challenges. Consequently, a wide variety of tangible applications and salient surface characteristics has been unveiled, for example, dust-free solar cells [13], non-wetting leathers [14], efficient and durable oil/water separators [15], anti-biofouling surfaces [16], anti-reflective surfaces [17], pump-free microfluidic and lab-on-chip devices [18], nanogenerators for energy harvesting [19], efficient water vapor condensation and enhanced heat transfer [20], as well as the stimuli-responsive surfaces [21], gas sensors [22], smart gating-based valves [23], and functionalized immunoassays [24].

This chapter shed a light on the bioinspired smart surfaces, and comprehensively elaborates the droplet dynamics. In between Section 1 (Introduction) and Section 5 (Conclusion), the main chapter breakdown provides three sections. By taking the examples of living surfaces, the classification of surface regimes, liquid droplet-surface interactions and the quantification criteria of wetting and dynamic regimes are included in Section 2. The major factors affecting the droplet dynamics have been unveiled in Section 3. The importance and role of droplet dynamics in emerging applications have been summarized in Section 4.

2. Classification of surface regimes and droplet-surface interactions

The surface regimes can be majorly categorized into wetting and dynamic regimes with reference to the interacting liquid droplets. In surface science, the living surfaces can provide four kinds of wetting regimes against water droplets, namely, hydrophilic, super hydrophilic, hydrophobic and superhydrophobic, as demonstrated in **Figure 2**. The droplet dynamics can be typically categorized into two main branches, namely, sticky and slippery regimes (**Figure 2**). The droplet dynamics is defined as the study of droplet growth, droplet mobility, droplet speed, droplet transport range and the underlying forces.

The droplet wetting regimes are conventionally quantified via equilibrium contact angle θ_e (°) that a liquid droplet makes with the surface, varying in a range of values (**Figure 1a**). It should be noted that, in several other studies, the inherent wettability of the pristine surfaces is denoted by the equilibrium contact angle θ_e (°), while the artificial wettability of the smart surfaces is denoted by the apparent contact angle θ_a (°). It means that the way to differentiate the pristine and smart surfaces can also be understood by the notation of contact angles. Guided by Young's model, θ_e (°) defines the static interaction of a stationary droplet with the flat topography of a horizontally-positioned solid substrate. Apparently, the solid substrate surfaces have a compact topography, but microscopically, the voids exist consisting of certain interspacing that affects the wetting interaction and create the philic (referring to hydrophilic and superhydrophilic) and phobic (referring to hydrophobic and superhydrophobic) regimes based on the extent of droplet penetration. For the philic regime, the micro-interspacing is supposed to exist into which droplet penetrates more intensely, and the opposite holds true for the phobic regime, i.e., nano-interspacing prohibits the droplet penetration. Thus, the surface topography affects the liquid droplet-surface interaction (LD-SI) that plays a decisive role regarding the droplet shape, i.e., with dominant micro-interspacing, the extent of droplet penetration is high, so the LD-SI area tends to be large with hemi to quarter-spherical droplet shape, specifying the philic regime in range of $0° \leq \theta_e \leq 90°$ [25–27]. While with dominant nano-interspacing, the LD-SI area is deemed to be small with hemi to full-spherical shape, which is called the phobic regime in the range of $90° \leq \theta_e \leq 180°$ [25–27]. The LD-SI can further be divided into two classes liquid droplet-dry surface interaction (LD-DSI) and liquid-droplet wet surface interaction (LD-WSI), as depicted in **Figure 2**. Depending on the LD-DSI and LD-WSI, the mechanism of droplet dynamics can be entirely changed as discussed below.

The droplet dynamic regimes can be quantified by the rolling/sliding angle α (°), which defines the dynamic interaction of a mobile droplet with the underlying tilted

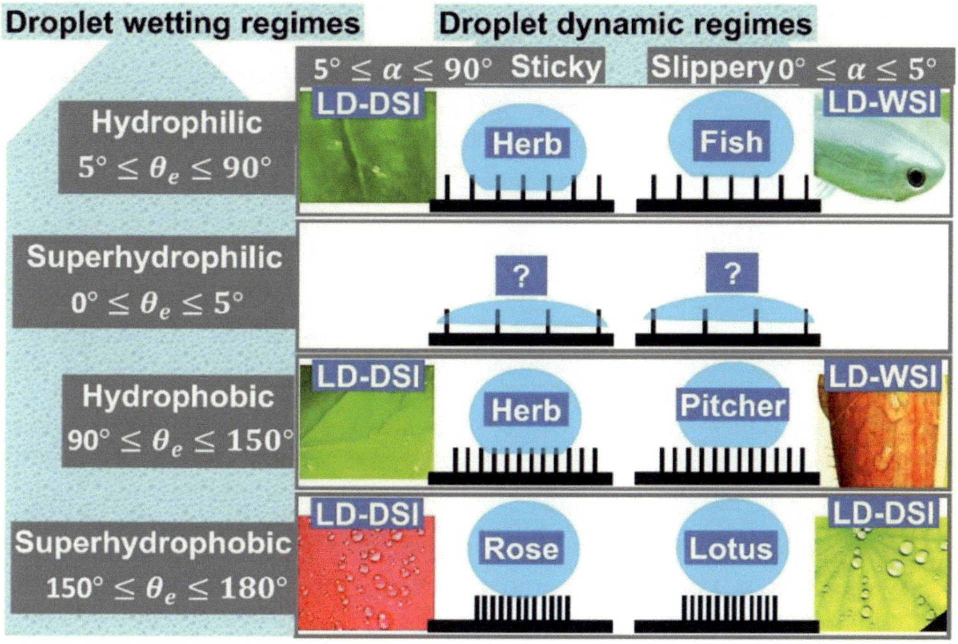

Figure 2.
Bioinspired classification of droplet wetting and droplet dynamic regimes. Liquid droplet-dry surface interaction (LD-DSI) and liquid droplet-wet surface interaction (LD-WSI) are depicted depending on the topographic states.

substrate. It should be noted that the rolling and sliding angles are specified just to distinguish the different surfaces due to dissimilar dynamic mechanisms depending on the droplet-surface interactions, e.g., the droplet is driven by the rolling mechanism on superhydrophobic surfaces due to LD-DSI (case of lotus leaf-inspired surfaces), while that on the slippery liquid-infused porous surfaces (SLIPSs), the sliding mechanism occurs due to LD-WSI (case of pitcher plant-inspired SLIPSs). But in both cases, α (°) is measured by the same method, i.e., the droplet is inserted onto the surface after which it is gradually tilted, and the angle is measured when the droplet starts moving. α is mainly helpful to demarcate the slippery and sticky regimes, regardless of the dynamic mechanisms of the involved surfaces types (superhydrophobic surfaces or SLIPSs).

The slippery dynamic regime, where virtually no droplet adhesion/pinning is considered, allows the ease of droplet mobility with value of α varying in range of $0° \leq \alpha \leq 5°$ [28]; while the sticky dynamic regime, where the droplet adhesion/pinning is variably considered, may provide different mobility behaviors with a range of $5° \leq \alpha \leq 90°$. A low adhesion Wenzel state or slippery Wenzel state can be conceived with a range of $5° \leq \alpha \leq 50°$ where the droplet pinning is relatively stronger compared with that of slippery regime [29]. A high adhesion Wenzel state corresponds to the range of $50° \leq \alpha \leq 90°$ [30], inducing the highest pinning against the droplet. A full adhesion Wenzel state occurs when $\alpha = 90°$ where the droplet does not move [31]. Therefore, the traditional norm defining that the Wenzel state induces the complete sticky regime (i.e., the droplet does not move at all), can be negated henceforth based on the above categorization and experimental proofs.

3. Factors affecting the droplet dynamics

Droplet dynamics is a complicated phenomenon, which is drastically affected by several salient features such as surface roughness, ASI, SAL, surface position (horizontal and inclined), topographic structural alignment (isotropic, anisotropic and gradient), LD-SI and droplet shape, and stimuli. They are elaborated as follows:

3.1 Roughness, ASI and SAL

The droplet dynamics is a very complicated phenomenon that simultaneously takes the topographic structural geometry, structural arrangement, structural direction, size scale, wetting regime, ASI (axial structural interspacing) and SAL (structural apex layer) into account.

Based on the values of θ_e/θ_a (°) and α (°), the wetting and dynamics can be interlinked in the above-categorized regimes (philic, phobic, slippery and sticky), but this link is supposed to be weak, especially when studying the droplet dynamics. Tuning the structural geometry with size from micro-scale to nano-scale (roughness) [31] can result in the philic to phobic regime, meaning that θ_a (°) can be symmetrically increased from 0° to 180°, but α (°) cannot be obtained in this order, i.e., it may either increase or decrease instead.

With a hierarchical structural arrangement even in the phobic regime, the droplet may lie in sticky regime where the adhesion can cause partial or even no mobility. The first example in such a case is the rose petal effect where the wetting lies in the extreme superhydrophobic regime but droplet does not move even at the tilt angle of 90°, evidencing the full adhesion Wenzel state [32]. The second example includes the butterfly wings and rice leaves with a superhydrophobic wetting regime where the structural direction simultaneously builds the sticky (full adhesion Wenzel state) and slippery regimes [33]. In addition, the efficient droplet dynamics do not completely rely on the wetting regime (phobic or philic), i.e., rose petal and lotus leave both have a hierarchical structural arrangement and lie in an extreme superhydrophobic regime [34], but the former does not support droplet mobility, while the latter is the most effective for droplet mobility. Indeed, the axial structural interspacing (ASI) is a prevalently crucial parameter [32] and with ASI as small as possible, the droplet dynamics can be made efficient. In general, the nano-scaled ASI [35] (producing the slippery regime with efficient droplet dynamics) is relatively encouraged compared with microscopic ASI; nonetheless, this principle does not always hold true, because now it is likely to convert the sticky regime even with high micro-scaled ASI into the perfect slippery regime with help of artificial SAL as discussed above. Thus, we conclude with reasoning that ASI and SAL are the main drivers together with entrapped-air as the secondary driver in all kinds of structural voids, establishing the exclusive droplet wetting and dynamic regimes.

3.2 Horizontal surface position

Horizontal surface position refers to the droplet lying in the static state. Indeed, the contact line [36] is the key factor influencing the extent of droplet adhesion, especially in the horizontal surface position. Herein, the droplet dynamics can be inferred by slowly evaporating the droplet resting on the surface, which can also be named evaporative droplet dynamics, as shown in **Figure 3**. It follows the simple rule, that is, the contact line immobility implies the sticky regime that works under the constant contact radius (CCR) mode [37], while the contact line mobility (normally inward)

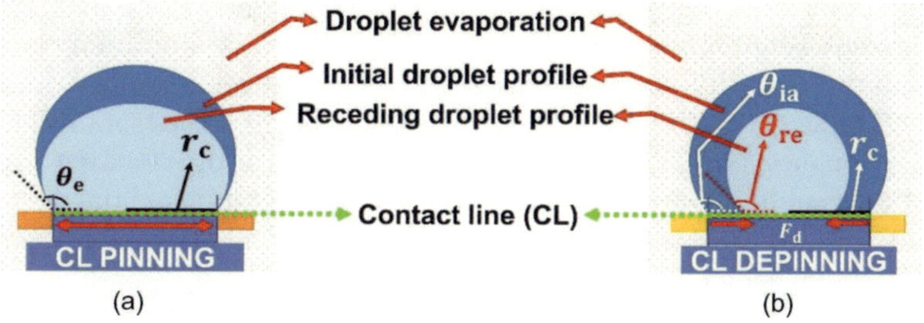

Figure 3.
Study of droplet dynamics under evaporation in horizontal surface position, depicting the (a) contact line (CL) pinning in CCR mode, and (b) depinning in CCA mode.

confirms the slippery regime working under the constant contact angle (CCA) mode [38]. In special cases, CCR and CCA mixed modes can also occur [39]. Quantitatively, the contact line is extracted in terms of the base diameter through experiments, and the depinning force F_d [40] can be measured.

In horizontal surface positions, the growth of droplets can also be carried out and controlled. For instance, the droplets tend to undergo frosting/icing/evaporation [41], and in such cases, the target is to delay or enhance the droplet growth rate.

In horizontal surface position, following factors also affect the droplet dynamics:

3.2.1 Wettability gradient

The specific arrangement of wetting regimes spanning from superhydrophobic to superhydrophilic can create the difference in wettability, known as the wettability gradient [42]. It plays an important role, providing the droplet mobility from a low wettability regime to a high wettability regime even when the surface is in horizontal position. This leads to self-propelled droplet mobility, i.e., no external energy in the form of gravity or stimulus is required [43]. Generally, with a high wettability gradient, the droplet dynamics are deemed to be efficient, and vice versa. An example of a wettability gradient can be seen in the beetle surface where the droplet dynamics play an important role in the water transport toward mouth [44].

3.2.2 Droplet impact

On the horizontal surface, the droplet impact can be obtained in two ways, i.e., by dropping the droplet from a certain height [45] or by coalescence of two or more droplets [46]. The droplet impact may exhibit a very unique phenomenon that strongly depends on the wetting regimes. While dropping the droplet, if the droplet rebounds several times, the slippery dynamics in superhydrophobic regime can be inferred; which is also known as droplet jumping, enabled by the conversion of energies [45, 46]. If the droplets do not rebound or show coalescence, the sticky dynamics are exhibited, which may lie in any wetting regime.

3.3 Inclined surface position

Depending on the wetting regime and droplet adhesion, the inclined mode allows the droplet movement under the effect of gravity, i.e., the gravitational force overcomes the under-lying resistive forces, which can be quantified by measuring the α (°).

The droplet does not move in the sticky regime, while it moves along the incline in the slippery regime, as exhibited in **Figure 4**. The droplet pinning/depinning is exhibited at the contact line (defined as the interface between the droplet and the surface topography) across the front/rear ridges. The contact angle at the front ridge is called the advancing angle θ_a, while at the rear ridge, it is known as the receding angle θ_r. The difference between the advancing angle and receding angle is called the contact angle hysteresis [31], i.e., the interfacial resistive forces attempting to prohibit the droplet mobility. In general, the smaller the CAH, the greater the droplet mobility.

The contact line plays the main role to overcome the interfacial resistive forces encountered by the droplets. The contact line tends to be discontinuous if the surface is in the slippery regime [33], requiring a smaller α that simply demonstrates the small interfacial resistive forces. Conversely, the contact line becomes continuous and stable (Wenzel state) in the sticky regime [33] due to which a larger α is indispensable.

In inclined surface position, further factors are important as discussed below:

3.3.1 Isotropic alignment

Isotropic is the surface feature in which the topographic structures are aligned equally, enabling droplet mobility in every direction. An example of such a case is the movement of a rain-drop on slippery surfaces such as taro leaf and lotus leaf [47].

3.3.2 Anisotropic alignment

Anisotropic is the surface feature in which the topographic structures are directionally-aligned, meaning that the droplet can only move along certain directions. For example, the rain droplet can move on the butterfly wings and rice leaves in only one direction [48].

3.3.3 Liquid transport modes

There are two most common liquid transport modes during the condensation process (conversion of the gaseous phase into the liquid phase), naming the filmwise

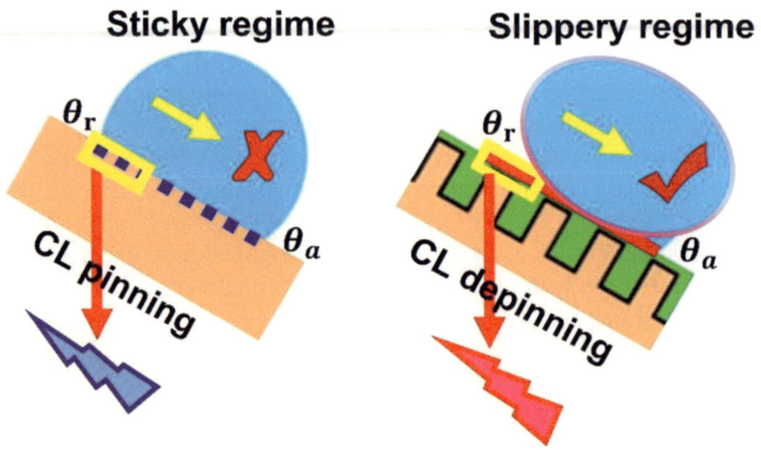

Figure 4.
Study of droplet dynamics in the inclined surface position.

mode and dropwise mode. They are studied normally on inclined surfaces. Each mode provides different droplet dynamics depending on the wetting regimes. In filmwise mode, droplets come into existence, soon after which the droplet coalescence (merging of small neighboring droplets to form a big droplet is called coalescence) starts occurring. Consequently, a thin liquid film is developed underneath that remains affixed to the surface topography, providing the under-layered path over which the bulk of liquid keeps on transporting. This happens due to the high surface energy that prompts the sticky regime where the filmwise mode occurs. Comparatively, the slippery regime, empowers the dropwise mode depending on the surface type underneath (i.e., either dry or wet). In the dropwise mode, the drops come into existence and move as soon as they attain the critical droplet size. The dry slippery hydrophobic/superhydrophobic surfaces give rise to the dropwise mode under the influence of rolling/jumping mechanisms [49] where the liquid drops tend to be transported either in the discrete pattern or coalesced pattern. While the wet slippery hydrophobic/hydrophilic surfaces support the dropwise mode enabled by the sliding mechanism [50] of droplets which can also either be in discrete or coalesced patterns.

3.4 LD-SI and droplet shape

The combined effect of LD-SI (droplet-surface interaction) and droplet shape may be considered a factor, however, droplet dynamics may not be fully predicted based on their combined effect. For example, the droplet usually adopts a full spherical shape (e.g., superhydrophobic lotus leaf and rose petals), hemispherical shape (e.g., hydrophobic cloverleaf) or quarter spherical shape (e.g., and hydrophilic herb) on LD-DSI, but some of them provide slippery dynamics, while some provide sticky dynamics. The droplet shape always tends to be a hemispherical shape (e.g., hydrophobic pitcher plant) or quarter spherical shape (e.g., hydrophilic beetle or fish surface) on LD-WSI, and they can all support slippery dynamics.

In the slippery regime, both LD-DSI and LD-WSI play a pivotal role in eventually deciding the two dynamic mechanisms or patterns of a mobile droplet (i.e., on the inclined surface). Owing to the underlying dry surface, the droplet tends to roll off, and the rolling mechanism is established as can be seen in the case of lotus leaf (superhydrophobic regime) and cloverleaf (hydrophobic regime). In contrast, because of the underlying wet surface, the droplet tends to slide-off over the liquid interface, and the sliding mechanism is produced as can be seen in case of pitcher plant and beetle surface.

3.5 Surface durability

Surface durability is of practical importance that can drastically affect the droplet dynamic regimes. The surface durability can be defined as the withstanding capability of bio-inspired surfaces deciding the active lifespan, and it can be realized with reference to the impact of several parameters and phenomena, including the solvents (i.e., neutral, acids, alkalis), weather conditions (i.e., temperature, pressure, airflow), operational conditions (i.e., condensation, evaporation, shearing, friction), etc. Depending on the applications and the types of involved surfaces, different parameters and phenomena can affect the surface durability up to different extents. In general, the low surface durability can be caused by many factors, naming a few, the poor growth of ETS/ITS, the poor attachment of ETS/ITS with the parent substrate (pristine topography), the poor adherence/infusion of SAL with the ETS/ITS, the poor chemical compatibility, amongst others. For example, the SLIPSs have shown practical challenges and low durability due to oil depletion through cloaking (the encapsulation of water droplets by the oil), evaporation (oil is evaporated at ambient or at high temperature) or physical shearing (the mobile droplets continuously induce friction at the droplet-surface interface and keeps on removing the oil with the passage of time) [30], finally resulting in low durability.

3.6 Stimuli

In practice, there are three well-defined major stimuli that affect droplet dynamics. First is the gravitational force (external stimulus) that comes into action when the surface is in an inclined position; Second is the active stimulus (external stimulus) provided by several sources including temperature, photon (light), pH, stress, mechanical vibration, gas, reactive solvents, magnetic field and electric potential [51]. Active stimulus can enable tunable droplet dynamics. For example, tuning the movement of a droplet on the paraffin-infused surfaces by changing the electric potential [52] or temperature [30]. It should however be noted that the tunable droplet dynamics are hard to be found on living surfaces. Third is the self-stimulus that is enabled by the wettability gradients. The various stimuli of droplet dynamics can co-exist depending on the wetting regimes and surface positions.

4. Droplet dynamics-lab scale and industrial applications

Based on the intended applications of non-living surfaces, the droplet dynamic regimes (whether sticky or slippery) are vital, e.g., the corrosion-combating surfaces [53] or anti-icing surfaces [54] are intensively required where the main target is to prohibit the contact between the water and surface [55], thus the sticky or moderately slippery dynamics are favorable. While in other cases such as water vapor condensation [56] and vehicle transportation through the water [57], the main target is the quick transport of water droplets on the surface and the drag reduction on the object, respectively, both necessitating the slippery dynamic regime. Therefore, droplet dynamic regimes are mainly decided by the applications. Generally, there are many applications, but a few emerging applications of droplet dynamics are presented underneath.

4.1 Droplet manipulation

Droplet manipulation can be defined as the control of droplets through various design strategies of under-lying surfaces as well as droplets, such as rendering the droplet mobile, immobile, directional movements, static growth, etc. A few case studies are listed underneath.

4.1.1 Droplet guided-tracks

Guided tracks are the directional pathways where the droplet can move with great ease, demonstrating the efficient droplet dynamics. Following the bioinspired mimicking, guided tracks can be made in various shapes, such as straight, inclined, s-shape, and zig-zag amongst others. The example is given in **Figure 5** [58], where the s-shaped track is depicted. The under-lying surface owns a slippery superhydrophobic regime, which has been precisely carved. Most importantly, the

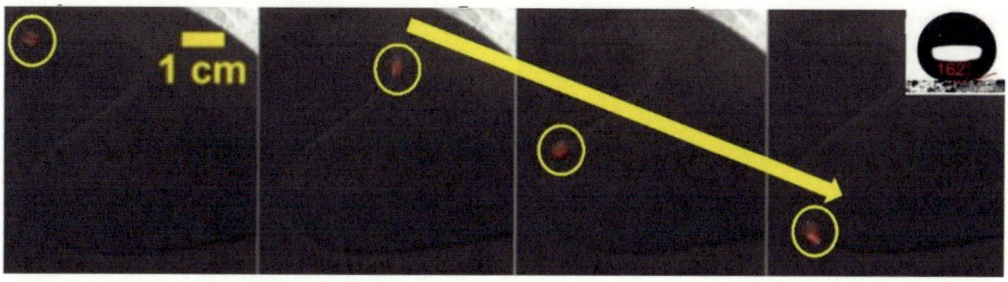

Figure 5.
Droplet transport along the guiding track [58].

carved region and the droplet volume need to be effectively controlled so that the droplet cannot stay out of the track.

4.1.2 Droplet segmentation/merging

Droplet segmentation/merging can be carried out in several ways, i.e., by making the segmented pathways over which the droplets can move or by merged pathways helping droplets come together and coalesced. This can be controlled by the bioinspired surface modifications as well as by the geometry modifications. For example, by modifying the surface of a common steel-made blade [58], the superhydrophobic regime can be obtained, which is then helpful to segment a big-sized droplet into two smaller droplets, as depicted in **Figure 6**.

4.1.3 Stimuli-responsive manipulation

As described above, various stimuli can affect the droplet mobility and droplet transport range. The droplets can be either oil or water. On the inclined phase change slippery liquid-infused porous surface (PC-SLIPSs) [30], the droplet mobility has been achieved in solid, mush and liquid phases under the temperature-stimulus which is also known as thermo-responsiveness, as shown in **Figure 7a**. The wetting regime is hydrophobic, while the dynamic regimes are influenced by the phases and temperature. The solid phase provides the low adhesion Wenzel (LAW) state. Particularly, at the melting temperature of PC-SLIPSs, the droplet mobility suddenly changes from high adhesion Wenzel (HAW) state into a slippery state. The oil-droplet mobility has been realized underwater by using the electric stimulus [59], depicting the controllable dynamics, as depicted in **Figure 7b**.

4.1.4 Self-propelled droplet manipulation

The droplet manipulation can be achieved without any external energy drivers or stimuli. The wettability gradient is the main driver, stimulating the self-mobility from low wettability toward high wettability. The example of steam condensation [60] is demonstrated (**Figure 8a**) where the liquid droplets are generated and move from left to right along the horizontal wires, leading to the droplet coalescence along the way. Eventually, droplet shedding occurs when the critical size of droplet has been achieved.

By means of the wettability gradient, the uphill droplet manipulation (**Figure 8b**) can also be achieved [43]. For example, the droplet can move upward on different

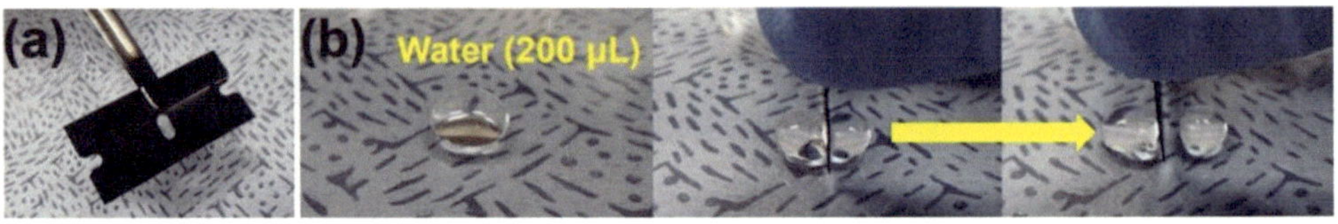

Figure 6.
Droplet segmentation via superhydrophobic blade [58].

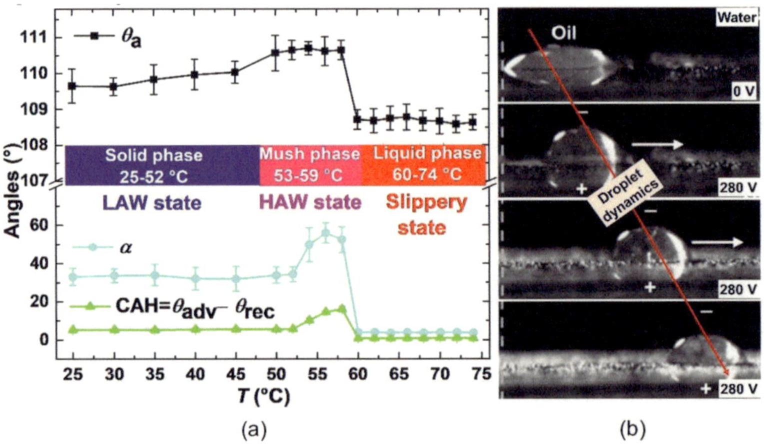

Figure 7.
(a) Thermo-responsive water droplet manipulation [30], (b) electro-responsive underwater oil droplet mobility [59].

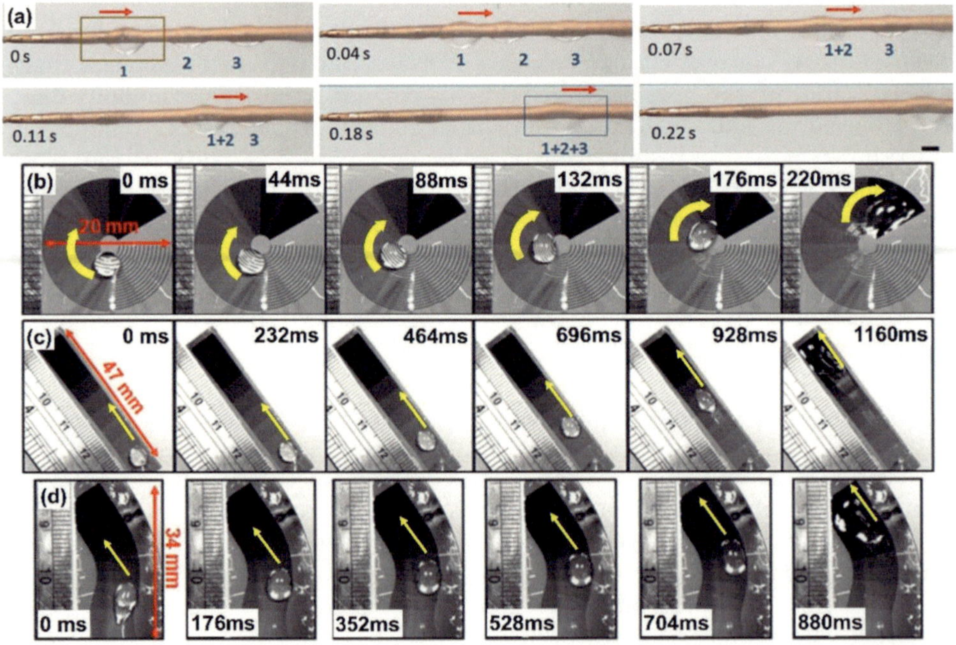

Figure 8.
Wettability gradient-induced self-propelled (a) droplet merging and shedding [60], as well as droplet uphill movement [43].

tracks, including circular geometry, straight geometry and s-shaped geometry. The droplet shape changes from the start until the end while covering the different transport ranges, depicting the various wetting regimes which are the characteristics of the wettability gradient.

4.2 Self-cleaning

When the mobile droplets move on the surfaces, they can carry the dust and dirt particles with them, which is named the self-cleaning effect. Therefore, the slippery dynamic regime is the key, that can be executed either on dry or wet slippery surfaces. For example, the droplet moves on the phase change material-based superhydrophobic surface, carrying the dust away as can be seen in **Figure 9** [61].

4.3 Chemical reactor and circuit controller

The mixed slippery and sticky regimes of the droplet dynamics can help build chemical reactors and electric circuit controllers. A dual stimuli-responsive SLIPSs have been presented [62], influenced by the temperature and force. A droplet (8 μL) of sodium hydroxide (NaOH) can be made mobile under the optimum effect of temperature and strain-induced force, letting it mix with the phenolphthalein droplet where the chemical reaction takes place, as exhibited in **Figure 10a**. Likewise, under the optimum effect of temperature and strain-induced force, the sliding and pinning of NaOH droplet can switch off and on the lamp, exhibiting the feasible electric controller (**Figure 10b**).

4.4 Condensation-water harvesting and heat transfer

The droplet dynamics play the most influential role in the condensation, where the filmwise mode and dropwise mode are considered. By means of condensation, water harvesting and heat transfer applications can be realized. In both of them, the dropwise mode is of great interest, providing the high droplet nucleation and ease of droplet transport which can be achieved through slippery dynamic regimes. For example, the dropwise mode has been demonstrated in the superhydrophobic regime and hydrophilic regime, enabling the fast droplet transport suitable for the efficient water harvesting (**Figure 11(a** and **b)**) [63]. Likewise, the heat transfer (**Figure 11c**)

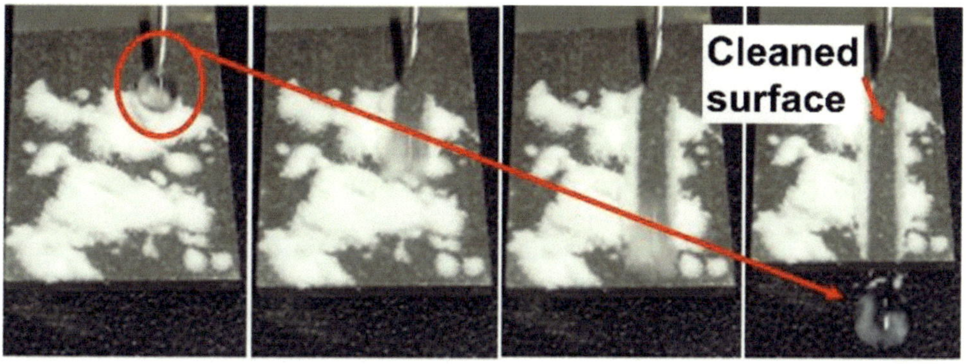

Figure 9.
Self-cleaning via mobile droplets [61].

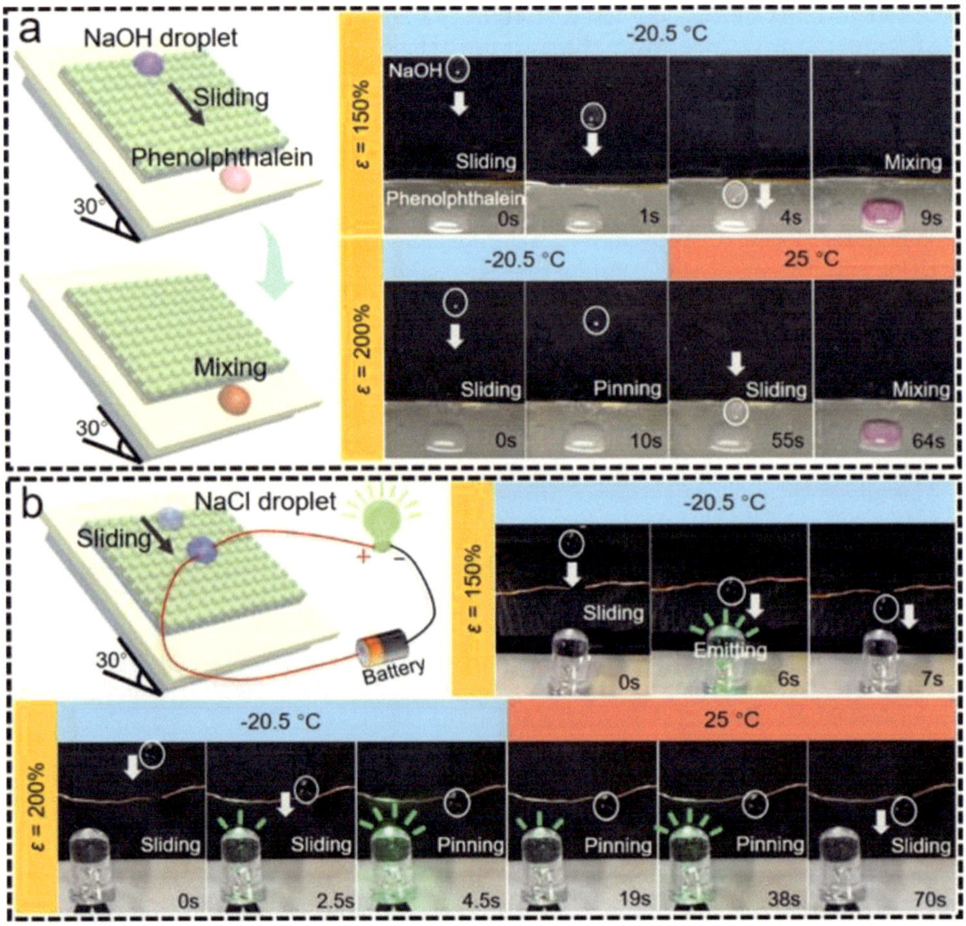

Figure 10.
Droplet sliding/pinning helping build (a) chemical reactor [62] and (b) electric circuit controller [62].

has been obtained higher in the slippery hydrophobic regime of phase change slippery liquid-infused porous surfaces (PC-SLIPSs) and slippery superhydrophobic regime compared to that in the sticky hydrophilic regime of pristine copper plate [20]. It should be particularly noted that efficient droplet dynamics are necessary during condensation, i.e., the dropwise mode may also underperform if the droplet shedding is slow as concluded in Ref. [20].

4.5 Oil/water separation

The oil/water separators are prepared by developing smart coatings on the porous networks. Depending on the wetting regimes of the separator, one phase is blocked, while the other is permitted through the porous network. Sticky and slippery regimes of droplet dynamics are crucial for oil/water separators; however, it should be particularly noted that these regimes may be insignificant during the separation mechanisms of vertically-aligned separators. It is because the blocking of one phase is not driven by the sticky dynamic regime, as well as the permitting of the other phase is not driven by the slippery dynamics. Instead, the separation mechanism is driven by the positive and negative capillary effects of two phases, inducing physical absorption

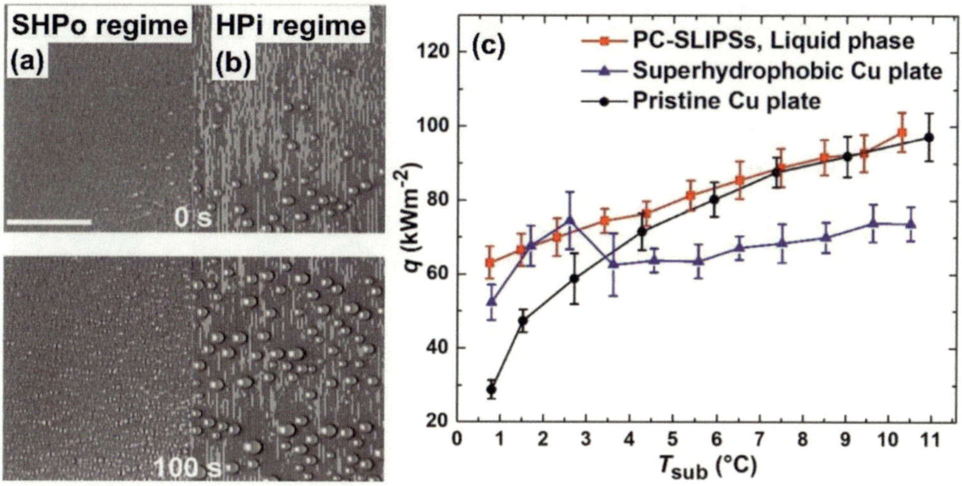

Figure 11.
Water harvesting under dropwise mode in (a) superhydrophobic (SHPo) regime and (b) hydrophilic (HPi) regime [63], (c) steam condensation heat transfer on pristine copper plate, superhydrophobic copper plate and PC-SLIPSs [20]. Scale bar, 5 mm.

and physical adhesion, as shown in **Figure 12a** [15]. Thus, in vertically-aligned separators, the sticky and slippery dynamics put forward a positive contribution in other ways. For example, the low sticky regime or moderately-high sticky regime is relatively effective, which can create easiness in the recovery process of separation medium after the oil/water separation. The self-recovery is important because the smart coatings can retain either oil or water during the separation process. With low sticky dynamics, the retained oil or water can easily evaporate, providing efficient self-recovery, as depicted in **Figure 12a**. The slippery dynamics can also play a vital role before or after the separation process, for example, a superhydrophobic-oleophilic copper mesh has been presented which is an effective oil/water separator. The slippery dynamics are of great significance to resolve the cleaning challenges of the separators (**Figure 12b**) [64].

4.6 Spraying

The spraying of different liquids is encountered almost everywhere, for example, spray of water, fuels, solvents, paints and perfumes, amongst others. In the spraying process, a multitude of droplets, consisting of various sizes, are generated. The droplet dynamics in the spray process can be affected by many factors, naming a few, the droplet type (i.e., oil, fuel, water, acids, alkalis, etc.), the droplet traveling speed, atomizer distance, and the surfaces position (horizontal, vertical or inclined). Upon impact with the surfaces, different droplets can behave differently depending on the wetting regimes and surface roughness. The impact of fuel droplets and surface roughness have been studied during the fuel spray in combustion engines [65]. It has been concluded that the various wetting regimes (depicted by the two values of surface roughness of $R_a \sim 2.5$ and 7.7) can produce various levels of fuel film thickness, depicting the unique film development mechanism (**Figure 13**). It shows that the droplet dynamics and the wetting regimes together play a pivotal role during the spraying, urging deeper studies in this field especially addressing the contact angles, sliding angles and contact angle hysteresis, etc.

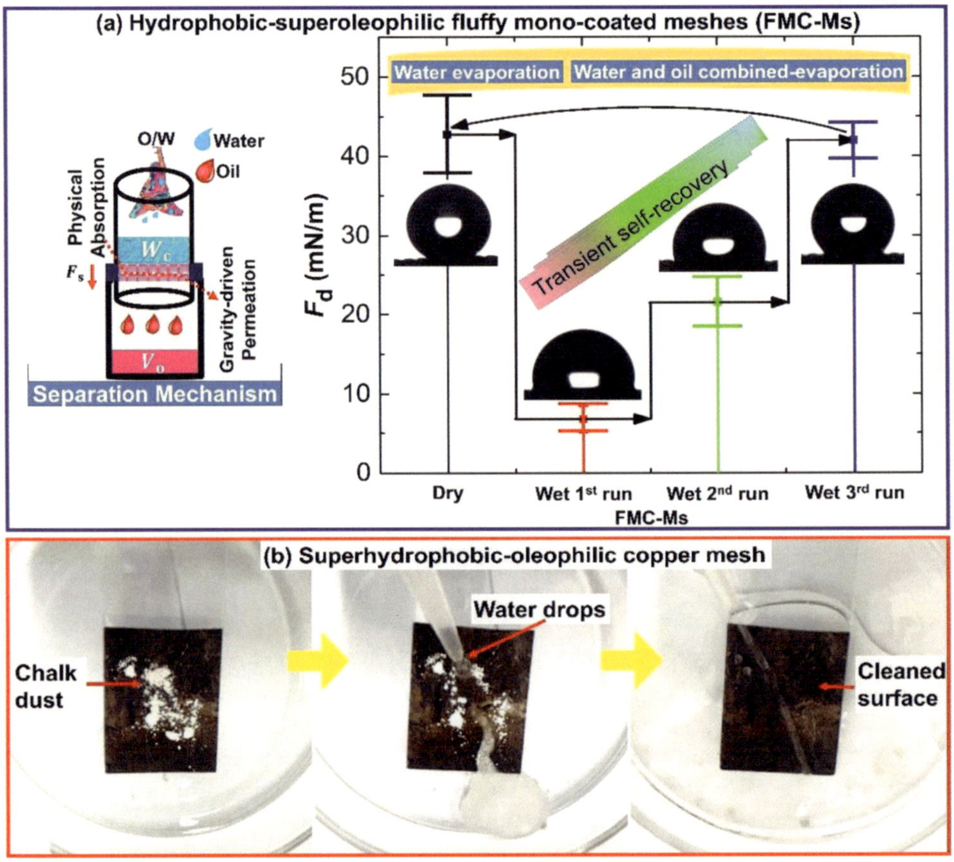

Figure 12.
Oil/water separators. (a) Low sticky dynamics depict the ease of self-recovery [15], (b) self-cleaning of separator [64].

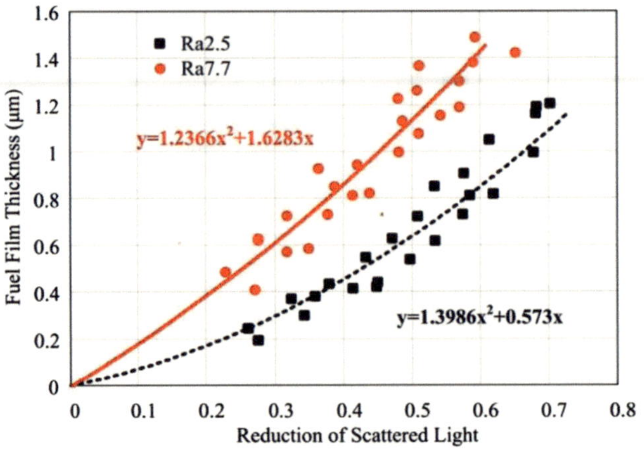

Figure 13.
Spray process of fuel droplets onto the horizontally-positioned glass slides having different roughness (R_a) values [65].

5. Conclusion

Bioinspired smart surfaces have been reported, with a special focus on the droplet dynamics. The surface regimes have been fundamentally classified into four droplet wetting and two droplet dynamic regimes. The droplet dynamics is defined as the study of the droplet growth, droplet mobility, droplet speed, droplet transport range and the underlying forces. The droplet dynamics is of great interest to a wide variety of scientific areas. With help of bioinspired smart surfaces, the droplet dynamics, either in sticky or slippery regimes, have been greatly improved. There are many factors that affect the droplet dynamics, that is, it is hard to rely on a single factor. In particular, the specific dynamic regime and the influential factors need to be modified depending on the applications, i.e., some applications may need the sticky regime, some may need the slippery regime, and some may need the co-existence of both regimes.

This chapter implies that the droplet dynamics are potentially significant, however further efforts are necessary to sustain the efficient droplet dynamics. It indeed depends on the chemical and mechanical strength of the droplet wetting regimes of the under-lying bioinspired surfaces, i.e., the surface durability. All bioinspired surfaces (superhydrophobic, SLIPSs, etc.), as presented so far, suffer from the bottlenecks corresponding to the durability, rendering the droplet dynamics inefficient up to various levels. It is therefore recommended to use alternative materials (e.g., solid slippery materials such as waxy and non-waxy phase change materials) and advanced surface engineering approaches to enhance the durability of the bioinspired surfaces that can sustain the droplet dynamics during real-time applications. Briefly, the efficient droplet dynamics can potentially enhance the performance of those systems intending to minimize the global energy crisis (e.g., less drag-inducing pipes can help save energy during the pumping process, slippery surfaces of photovoltaics can prohibit/postpone the ice accretion as well as avoid the formation of dust layers), save the operational time (e.g., condensers can provide high output in less time based on the high droplet transport) and provide the high throughput (e.g., microfluidic devices).

Acknowledgements

The support provided by the IntechOpen, as well as by the Postdoctoral Program of Southeast University are highly acknowledged.

Author details

Raza Gulfam
School of Energy and Environment, Southeast University, Nanjing, China

*Address all correspondence to: gulfamrazahaidery@hotmail.com;
gulfam@seu.edu.cn

References

[1] Yong J, Chen F, Yang Q, Jiang Z, Hous X. A review of femtosecond-laser-induced underwater superoleophobic surfaces. Advanced Materials Interfaces. 2018;**5**:1701370

[2] Cao M, Jiang L. Superwettability integration: Concepts, design and applications. Surface Innovations. 2016;**4**:180-194

[3] Wen RF, Ma X. Advances in dropwise condensation: dancing droplets. In: Pham P, Goel P, Kumar S, Yadav K, editors. 21st Century Surface Science-a Handbook. UK: IntechOpen; 2020. p. 92689

[4] Kerstin K, Wilhelm B. Superhydrophobic and superhydrophilic plant surfaces: An inspiration for biomimetic materials. Philosophical Transactions of the Royal Society A. 2009;**367**:1487-1509

[5] Atsushi T, Takaharu T. Enhancement of condensation heat transfer on a microstructured surface with wettability gradient. International of Journal of Heat and Mass Transfer. 2020;**156**:119839

[6] Zhang P, Lv FY. A review of the recent advances in superhydrophobic surfaces and the emerging energy-related applications. Energy. 2015;**82**:1068-1087

[7] Zhongzhen W, Liangliang L, Shunning L. Tuning superhydrophobic materials with negative surface energy domains. Research. 2019;**2019**:1391804

[8] Weijian L, Mingyong C, Xiao L, et al. Wettability transition modes of aluminum surfaces with various micro/nanostructures produced by a femtosecond laser. Journal of Laser Applications. 2019;**31**:022503

[9] Scholz I, Bückins M, Dolge L, Erlinghagen T, Weth A, Hischen F, et al. Slippery surfaces of pitcher plants: Nepenthes wax crystals minimize insect attachment via microscopic surface roughness. Journal of Experimental Biology. 2010;**213**:1115-1125

[10] Xu S, Wang Q, Wang N. Chemical fabrication strategies for achieving bioinspired superhydrophobic surfaces with micro and nanostructures: A review. Advanced Engineering Materials. 2021;**23**:2001083

[11] Wen G, Guo ZG, Liu W. Biomimetic polymeric superhydrophobic surfaces and nanostructures: From fabrication to applications. Nanoscale. 2017;**9**:3338-3366

[12] Feng J, Tuominen MT, Rothstein JP. Hierarchical superhydrophobic surfaces fabricated by dual-scale electron-beam-lithography with well-ordered secondary nanostructures. Advanced Functional Materials. 2011;**21**:3715-3722

[13] Syafiq A, Pandey AK, Adzman NN, Rahim NA. Advances in approaches and methods for self-cleaning of solar photovoltaic panels. Solar Energy. 2018;**162**:597-619

[14] Gurera D, Bhushan B. Fabrication of bioinspired superliquiphobic synthetic leather with self-cleaning and low adhesion. Colloids and Surfaces, A: Physiochemical and Engineering Aspects. 2018;**545**:130-137

[15] Gulfam R, Zhang P. Fabrication and characterization of fluffy mono-coated copper meshes and their applications for oil/water separation. Colloids and Surfaces A: Physicochemical and Engineering Aspects. 2021;**625**:126883

[16] Wang P, Zhang D, Sun S, Li T, Sun Y. Fabrication of slippery lubricant-infused porous surface with high underwater transparency for the control of marine biofouling. ACS Applied Materials and Interfaces. 2017;**9**:972-982

[17] Wang SD, Shu YY. Superhydrophobic antireflective coating with high transmittance. Journal of Coatings Technology and Research. 2013;**10**:527-535

[18] Shang L, Yu Y, Gao W, Wang Y, Qu L, Zhao Z, et al. Bio-inspired anisotropic wettability surfaces from dynamic ferrofluid assembled templates. Advanced Functional Materials. 2018;**28**:1705802

[19] Wang Y, Gao S, Xu W, Wang Z. Nanogenerators with superwetting surfaces for harvesting water/liquid energy. Advanced Functional Materials. 2020;**30**:1908252

[20] Gulfam R, Huang TE, Lv C, Orejon D, Zhang P. Condensation heat transfer on phase change slippery liquid-infused porous surfaces. International Journal of Heat and Mass Transfer. 2022;**185**:122384

[21] Lou X, Huang Y, Yang X, Zhu H, Heng L, Xia F. External stimuli responsive liquid-infused surfaces switching between slippery and nonslippery states: Fabrications and applications. Advanced Functional Materials. 2020;**30**:1901130

[22] Liu Y, Wang X, Fei B, Hu H, Lai C, Xin JH. Bioinspired, stimuli-responsive, multifunctional superhydrophobic surface with directional wetting, adhesion, and transport of water. Advanced Functional Materials. 2015;**25**:5047-5056

[23] Sun Z, Cao Z, Li Y, Zhang Q, Zhang X, Qian J, et al. Switchable smart porous surface for controllable liquid transportation. Materials Horizons. 2022;**9**:780-790

[24] Fang L, Zhang J, Chen Y, Liu S, Chen Q, Ke A, et al. High-resolution patterned functionalization of slippery "liquid-like" brush surfaces via microdroplet-confined growth of multifunctional polydopamine arrays. Advanced Functional Materials. 2021;**31**:2100447

[25] Mortazavi V, Khonsari MM. On the degradation of superhydrophobic surfaces: A review. Wear. 2017; **372-373**:145-157

[26] Zhan S, Pan Y, Gao ZF, Lou X, Xia F. Biological and chemical sensing applications based on special wettable surfaces. Trends in Analytical Chemistry. 2018;**108**:183-194

[27] Webb HK, Crawford RJ, Ivanova EP. Wettability of natural superhydrophobic surfaces. Advances in Colloid and Interface Science. 2014;**210**:58-64

[28] Prakash CGJ, Prasanth R. Recent trends in fabrication of nepenthes inspired SLIPSs: Design strategies for self-healing efficient anti-icing surfaces. Surface and Interfaces. 2020;**21**:100678

[29] Dai X, Stogin BB, Yang S, Wong TS. Slippery Wenzel state. ACS Nano. 2015;**9**:9260-9267

[30] Gulfam R, Orejon D, Choi CH, Zhang P. Phase-change slippery liquid-infused porous surfaces with thermo-responsive wetting and shedding states. ACS Applied Materials and Interfaces. 2020;**12**:34306-34316

[31] Cheng Z, Hou R, Du Y, Lai H, Fu K, Zhang N, et al. Designing heterogeneous chemical composition on hierarchical structured copper substrates for the fabrication of superhydrophobic

surfaces with controlled adhesion. ACS Applied Materials and Interfaces. 2013;**5**:8753-8760

[32] Feng L, Zhang Y, Xi J, Zhu Y, Wang N, Xia F, et al. Petal effect: A superhydrophobic state with high adhesive force. Langmuir. 2008;**24**:4114-4119

[33] Liu M, Zheng Y, Zhai J, Jiang L. Bioinspired super-antiwetting interfaces with special liquid-solid adhesion. Accounts of Chemical Research. 2010;**43**:368-377

[34] Jeevahan J, Chandrasekaran M, Joseph GB, Durairaj RB, Mageshwaran G. Superhydrophobic surfaces: A review on fundamentals, applications, and challenges. Journal of Coatings Technology and Research. 2018;**15**:231-250

[35] Wan Y, Zhang C, Zhang M, Xu J. Anti-condensation behavior of bamboo leaf surface (backside) and its bionic preparation. Materials Research Express. 2021;**8**:055002

[36] Wang H. From contact line structures to wetting dynamics. Langmuir. 2019;**35**:10233-10245

[37] Erbil HY, Glen M, Newton MI. Drop evaporation on solid surfaces: Constant contact angle mode. Langmuir. 2002;**18**:2636-2641

[38] Hu H, Larson RG. Evaporation of a sessile droplet on a substrate. The Journal of Physical Chemistry B. 2002;**106**:1334-1344

[39] Ramos SMM, Dias JF, Canut B. Drop evaporation on superhydrophobic PTFE surfaces driven by contact line dynamics. Journal of Colloid and Interface Science. 2015;**440**:133-139

[40] Wei X, Choi CH. From sticky to slippery droplets: Dynamics of contact line depinning on superhydrophobic surfaces. Physical Review Letters. 2012;**109**:024504

[41] Ewetola M, Aguilar RL, Pradas M. Control of droplet evaporation on smooth chemical patterns. Physical Review Fluids. 2021;**6**:033904

[42] Wu J, Yin K, Xiao S, Wu Z, Zhu Z, JA D, et al. Laser fabrication of bioinspired gradient surfaces for wettability applications. Advanced Materials Interfaces. 2021;**8**:2001610

[43] Liu C, Sun J, Li J, Xiang C, Che L, Wang Z, et al. Long-range spontaneous droplet self-propulsion on wettability gradient surfaces. Scientific Reports. 2017;**7**:7552

[44] Parker AR, Lawrence CR. Water capture by a desert beetle. Nature. 2001;**414**:33-34

[45] Weisensee PB, Tian J, Miljkovic N, King WP. Water droplet impact on elastic superhydrophobic surfaces. Scientific Reports. 2016;**6**:30328

[46] Wang K, Ma X, Chen F, Lan Z. Effect of a superhydrophobic surface structure on droplet jumping velocity. Langmuir. 2021;**37**:1779-1787

[47] Singh RA, Yoon ES. Biomimetics in tribology-recent developments. Journal of the Korean Physical Society. 2008;**52**:661-673

[48] Bixler GD, Bhushan B. Bioinspired rice leaf and butterfly wing surface structures combining shark skin and lotus effects. Soft Matter. 2012;**8**:11271-11284

[49] Wang R, Wu F, Xing D, Yu F, Gao X. Density maximization of

one-step electrodeposited copper nanocones and dropwise condensation heat-transfer performance evaluation. ACS Applied Materials and Interfaces. 2020;**12**:24512-24520

[50] McCerery R, Woodward J, Glen M, Winter K, Armstrong S, Orme BV. Slippery liquid-infused porous surfaces: The effect of oil on the water repellence of hydrophobic and superhydrophobic soils. European Journal of Soil Sciences. 2021;**72**:963-978

[51] Li Y, He L, Zhang X, Zhang N, Tian D. External-field-induced gradient wetting for controllable liquid transport: From movement on the surface to penetration into the surface. Advanced Materials. 2017;**29**:1703802

[52] Gao W, Wang J, Zhang X, Sun L, Chen Y, Zhao Y. Electric-tunable wettability on a paraffin-infused slippery pattern surface. Chemical Engineering Journal. 2020;**381**:122612

[53] Lee J, Shin S, Jiang Y, Jeong C, Stone HA, Choi CH. Oil-impregnated nanoporous oxide layer for corrosion protection with self-healing. Advanced Functional Materials. 2017;**27**: 1606040

[54] Vercillo V, Tonnicchia S, Romano JM, Girón AG, Morales AIA, Alamri S, et al. Design rules for laser-treated icephobic metallic surfaces for aeronautic applications. Advanced Functional Materials. 2020;**30**:1910268

[55] Chatterjee R, Beysens D, Anand S. Delaying ice and frost formation using phase-switching liquids. Advanced Materials. 2019;**31**:1807812

[56] Deng Z, Zhang CB, Shen C, Cao J, Chen Y. Self-propelled dropwise condensation on a gradient surface. International Journal of Heat and Mass Transfer. 2017;**114**:419-429

[57] Dong H, Cheng M, Zhang Y, Wei H, Shi F. Extraordinary drag-reducing effect of a superhydrophobic coating on a macroscopic model ship at high speed. Journal of Materials Chemistry A. 2013;**1**:5886-5891

[58] Seo K, Kim M, Kim DH. Candle-based process for creating a stable superhydrophobic surface. Carbon. 2013;**68**:583-596

[59] Tian D, He L, Zhang N, Zheng X, Dou Y, Zhang X, et al. Electric field and gradient microstructure for cooperative driving of directional motion of underwater oil droplets. Advanced Functional Materials. 2016;**26**:7986-7992

[60] Xu T, Lin Y, Zhang M, Shi W, Zheng Y. High-efficiency fog collector: Water unidirectional transport on heterogeneous rough conical wires. ACS Nano. 2016;**10**:10681-10688

[61] Fan Y, He Y, Luo P, Chen X, Liu B. A facile electrodeposition process to fabricate corrosion-resistant superhydrophobic surface on carbon steel. Applied Surface Science. 2016;**368**:435-442

[62] Wu S, Liu L, Zhu S, Xiao Y. Smart control for water droplets on temperature and force dual-responsive slippery surfaces. Langmuir. 2021;**37**:578-584

[63] Dai X, Sun N, Nielsen SO, Stogin BB, Wang J, Yang S, et al. Hydrophilic directional slippery rough surfaces for water harvesting. Science. Advances. 2018;**4**:eaaq0919

[64] Zhang Z, Pengyu Z, Gao Y, Yun J. Fabrication of superhydrophobic copper meshes via simply soaking for oil/ water separation. Colloids and Surfaces A: Physicochemical and Engineering Aspects. 2022;**642**:128648

[65] Luo H, Uchitomi S, Nishida K, Ogata Y, Zhang W, Fujikawa T. Experimental investigation on fuel film formation by spray impingement on flat walls with different surface roughness. Atomization and Sprays. 2017;**27**:611-628

Chapter 6

Internal Flow and Spray Dynamics of Multi-Hole Nozzle

PengBo Dong

Abstract

Faced with dual challenges of "carbon neutral" and emission control, fossil fuel-based internal combustion engines need to explore new ways and technical paths to reduce harmful emissions and Carbon dioxide emissions simultaneously. Fuel injection process is playing a significant role not only in traditional engines but also in new low/zero carbon engines. Multi-hole nozzles have a wide range of applications in the fuel supply system. While the accepted spray study work and jet break-up models are usually developed under the quasi-steady-state of fuel injection by a single-hole nozzle. There are rare models that can describe the whole break-up processes of multi-hole nozzle spray, including complex internal flow factors, plume interaction, and the effect of start/end of injection. In this chapter, characteristics of spray morphology, evolution processes, and evaporation characteristics, emerging from the practical diesel multi-hole nozzles, were discussed and analyzed during the transient injection processes in detail. Moreover, the relationship between multi-hole nozzle internal flow properties and the corresponding spray behaviors was investigated by numerical simulation method systematically. Therefore, multi-hole spray modeling processes under engine operating conditions and the optimized design of diesel multi-hole nozzles are expected to get some benefits and clues from the current results.

Keywords: fuel injection, nozzle internal flow, cavitation, near-field spray, multi-hole nozzle, spray diagnostics, diesel engine

1. Introduction

The research on fuel sprays of Internal Combustion Engines has been drawing attention for last century. As for Diesel Engines, spray evolution, which includes the jet breakup, fuel atomization, air entrainment, and mixture formation processes, is regarded as one of the determinants of engine performance and emission formation. Furthermore, nozzle geometry can directly affect the characteristics of the complicated internal flow patterns inside the nozzles, such as the cavitated turbulence. As a result, in order to improve the quality of the atomization and mixture homogeneity, researchers have spared no effort to investigate the internal flow [1–3] and spray behaviors [4–6], and many fundamental and classic theories were formed over the past decades. Originally, due to the simple structure and the easiness of applying diagnostic techniques and arranging instruments surrounding the nozzle and the spray plume, single-hole nozzle was widely used in fundamental research.

Consequently, spray and combustion models, which are adopted in the numerical simulation study are usually developed from the experimental results of the single-hole nozzle spray.

On the other hand, it is known that multi-hole nozzles, which can emerge several spray plumes simultaneously, are generally applied in the engineering field of practical Diesel Engines. In this case, the orifices are normally aligned around the sac symmetrically and located off-axis to the nozzle axis. As a result, the optical access to the complex configuration and the mutual interference between the spray plumes create a difficulty for scholars to explore the internal flow and spray characteristics of multi-hole nozzles.

In recent years, there is a trend that considerable efforts have been exerted to conduct an experimental and computational study about the internal flow and sprays of the multi-hole nozzles. Different kinds of verisimilar scaled-up or real-size multi-hole nozzle models were designed to reveal the flow pattern properties inside the nozzles and the initial emerging spray dynamics [7–12]. The application of the synchrotron x-ray sources with high energy pulses has also been extended to the study of fuel sprays during the last decade [13, 14]. At the same time, a series of studies using the three-dimensional computational fluid dynamics simulations were also conducted to link the experimental results to the numerical calculation [15–18]. Based on all the approaches mentioned above, the vortex flow and string-type cavitation inside the sac, the counter-rotating vortices, film-type and string-type cavitation inside the hole volume, the needle lateral oscillation effect, and the unstable spray behaviors were found in succession.

Moreover, smaller hole diameter and more holes, accompanied by higher injection pressure are becoming prevalent in dealing with the strict emission regulations and economic demand [19, 20], because it is proven to be able to generate better fuel atomization and more homogeneous fuel/air mixture [21, 22]. In the conventional study, hole diameter is usually larger than 0.10 mm. Even though a few researchers applied the micro-hole nozzle to conduct the fundamental experimental study about the mixture formation and combustion processes, mainly single-hole nozzles were used in their experiments. As a result, the limited nozzle types and experimental conditions hindered a more satisfactory situation in understanding the flow dynamics of the real multi-hole nozzles under practical operating conditions thoroughly. There is rare information about the spray properties of multi-hole nozzles with micro orifices in the current archives, and it is worthwhile to carry out a monographic study on this issue.

A comparison between the spray properties of the nozzles with different hole diameter is made in the present chapter to provide deep insights into the multi-hole nozzles sprays, which are formed in the real scenario of Diesel Engines, and sufficient analyses about the effect of the hole diameter were conducted systematically as well. In addition, the CFD simulation results for these different nozzle configurations were also presented in this chapter with the aim of correlating the observed spray behaviors to the internal flow properties inside the nozzles.

2. Methodology

2.1 Experimental apparatus, procedures, and conditions for spray observation

High-speed video camera observations based on the Mie scattering were made for the sprays injected by the multi-hole injectors, and the specific experimental apparatus is shown in **Figure 1**. **Figure 1(a)** presents the global experimental arrangement, including the high-pressure chamber, fuel injection system, and optical system. A common rail injection system could generate the injection pressure up to 220 MPa, and high-pressure constant volume vessels with optically accessible quartz windows were employed to create the high-pressure ambient environment of Diesel fuel injection. A delay pulse generator (Stanford Inc., DG535) and an electronic control unit (ECU) were applied to control the image timing, injection time, and injection quantity. The optical path layout is shown in **Figure 1(a)**. A xenon lamp (USHIO Corp., SX-UID501XAMQ) and two high-speed video cameras (Nac MEMRECAM HX-3, Photron FASTCAM-APX RS) were used to record the fuel injection processes.

When it comes to the installation of the multi-hole injectors, the detailed information about the specially designed chamber head is shown in **Figure 1(b)**. It is well known that the observation of the multi-hole nozzle sprays is difficult due to the conical structure formed by the plumes. In this study, the multi-hole injector was installed into this specialized chamber cover obliquely, as shown in **Figure 1(c)**, to prevent spray interference. An appropriate angle between the axis of the multi-hole nozzle and the horizontal plane was designed to maintain that one of the spray plumes could be observed as vertically as possible.

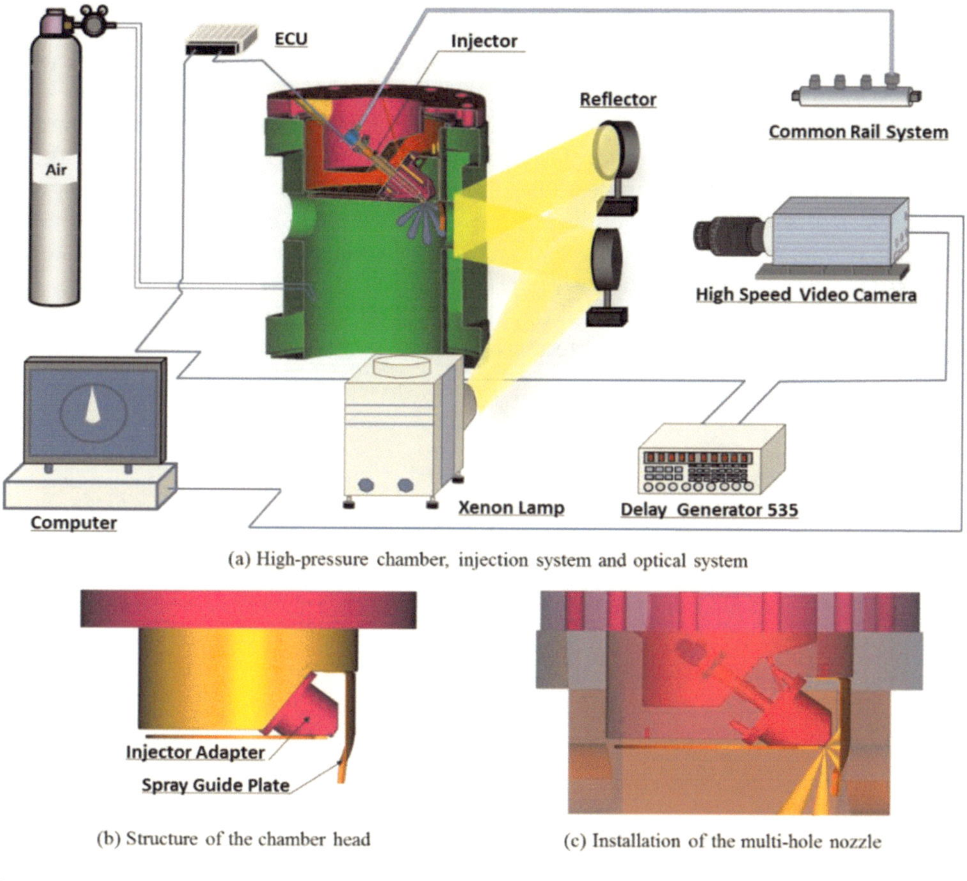

(a) High-pressure chamber, injection system and optical system

(b) Structure of the chamber head

(c) Installation of the multi-hole nozzle

Figure 1.
Experimental apparatus.

The experimental apparatus for the injection rate measurement are presented in **Figure 2**.

In this study, the Bocsh rate of injection meter [23] was applied. The experimental conditions corresponded to the spray observation experiments, which will be introduced in the upcoming sections. The timing of the start of injection was determined by comparing the injection pause signal with the injection rate curve signal recorded in the oscilloscope.

Figure 3 shows the schematics of the nozzles applied in this study, the two multi-hole nozzles (10 holes) have the same configuration except for the hole diameter (D = 0.10 and 0.07 mm). The experimental conditions, which are shown in detail in **Table 1**, were determined in consideration of the real operation conditions of small Diesel engines.

The setting of cameras is also presented in **Table 1**, and the experimental measurement was conducted at least 10 times for each condition. The test fuel was the JIS #2 diesel. The injection quantity was 2 mm^3/hole. In order to keep the ambient gas density similar to that of the combustion conditions, for fundamental spray research of non-evaporation conditions, the ambient temperature and pressure were 300 K and 1.5 MPa, respectively.

2.2 Methods of image processing

The typical image processing processes and the definitions of spray properties in the current study are shown in **Figure 4**. The same processing method was applied to the spray images of different nozzles. The spray image taken under the baseline condition is shown here as an example. The central spray of each frame was characterized as the spray tip penetration (i.e., the maximum penetration distance of the spray, S), the corresponding angle of 100 times hole diameter length (i.e., the spray cone angle, θ_c), and the corresponding angle at the half point of S (i.e., the spray angle, θ_s). The parameters are plotted as a function of the time after start of injection and nozzles with different hole diameters.

The spray images were processed to calculate their properties by the following steps. First, each frame was converted to an effective image by subtracting the

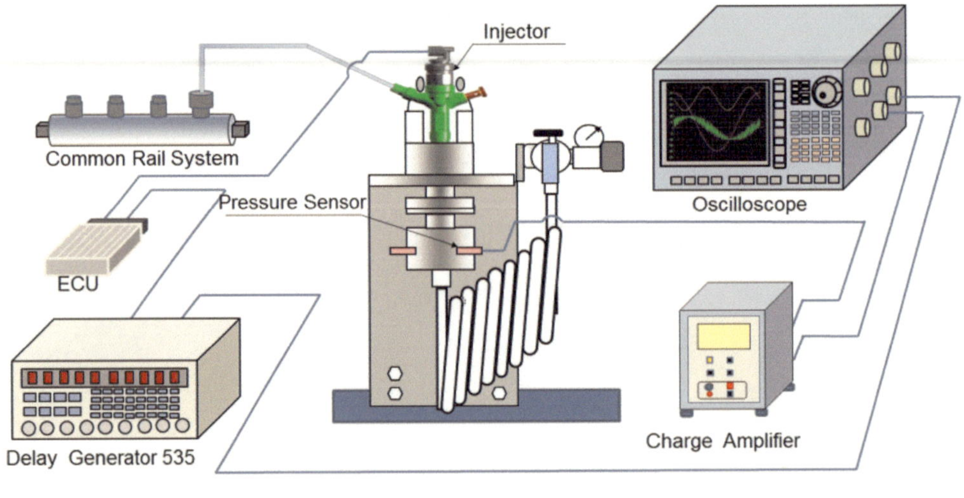

Figure 2.
Schematic of injection rate measurement experiment apparatus.

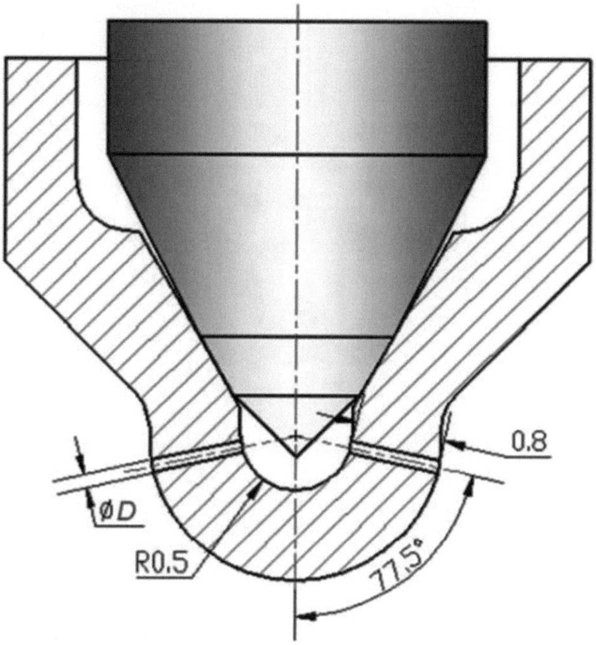

Figure 3.
Schematic of different nozzles applied in experiments.

background images taken without the spray injection. After that, the spray edge could be detected by using the binarization image, which is converted by a threshold algorithm. The colored edges of different recording times show that the spray contour and the temporal variations can be well captured spatially. The spray tip penetration of the central spray plume is determined by scanning the contours from the corresponding orifice point. In this way, the morphological algorithm could also be used to extract the parameters of spray cone angle and the spray angle. The near-field spray properties around the nozzle exit region were also captured by the same method introduced above.

One of the most important factors that can introduce uncertainty into the measurement results is the selection of the threshold. Compared with the single-hole nozzle, the hole-to-hole variation and the cycle-to-cycle variation of multi-hole nozzle sprays cause the image processing to be more sensitive to the algorithm. Hence, the assessment of the threshold value is necessary for the current study. Principles that are suitable for the threshold selection are summarized as (1) The threshold should ensure that the spray profiles are as similar to the raw images as possible; (2) The threshold should remove the background noises in the images; (3) The error should be within the cycle-to-cycle injection variations. As a result, the intensity threshold of 5 (the maximum intensity: 255) was selected in this study after a series of statistics.

2.3 Computational setting for internal flow study

In the current study, to aid the interpretation of the experimental results, the influences of the micro-hole diameter on the internal flow and cavitation characteristics have been numerically investigated by the commercial CFD Code FIRE Version

Injection Conditions	
Fuel (Diesel JIS#2)	Density (20°C,1 atm): ~830 kg/m^3 Boiling Point (1 atm): ~273°C Kinetic Viscosity (20°C,1 atm): ~3.86 (10^{-6}) m^2/s Cetane Index: ≥45 Sulfur: 10 ppm
Injector	Multi-Hole Nozzle
Injection Quantity Q_{inj} (mm^3/hole)	2.0
Rail Pressure P_{rail} (MPa)	120
Hole Diameter D (mm)	0.07 0.10*
Injection Duration: $\triangle t_{inj}$ (ms)	1.26 0.84*
Ambient Conditions	
Ambient Gas	Air
Ambient Pressure P_a (MPa)	1.5
Ambient Temperature T_a (K)	300
Ambient Density ρ_a (kg/m^3)	17.4
High-Speed Video Camera Conditions	
Recording Rate (fps)	10,000 100,000
Exposing Time (s)	1/10000 1/100000
Resolution	512 × 512,384 × 200

Baseline Condition.

Table 1.
Experimental conditions.

2017 (AVL). **Figure 5** shows the computational meshes of the multi-hole nozzle with ten holes, and only one-tenth of the entire volumetric domain was selected considering the geometric periodicity, symmetry, and calculation timing.

The specific settings for this computation have been listed in **Table 2**. The same setting was applied to the two different nozzle configurations to make comparisons between the internal flow patterns. The Reynolds Averaged Navier–Stokes Simulation (RANS) model and a four-equation k-ζ-f model developed from the standard k-ε model were adopted to simulate the turbulent flow. The k-ζ-f model introduces new transportation equations to describe the variable ζ which has a relationship with the turbulence viscosity. As a result, the property of anisotropic turbulence can be taken into the consideration. A multi-phase flow model was selected to approximate the fluid conditions inside the nozzles. Furthermore, a Linearized Rayleigh model [24] was used to express the cavitation bubble behaviors within the nozzle.

Validation was conducted by taking the experimental results published by Blessing et al. [25] as the criteria, in which the characteristics of nozzles and boundary conditions covered many features of the current study, and the turbulence and cavitation models were proved reasonable before further computational studies. Moreover, the effect of mesh size was also taken into the consideration by making comparison between the different computational results from the meshes with different mesh

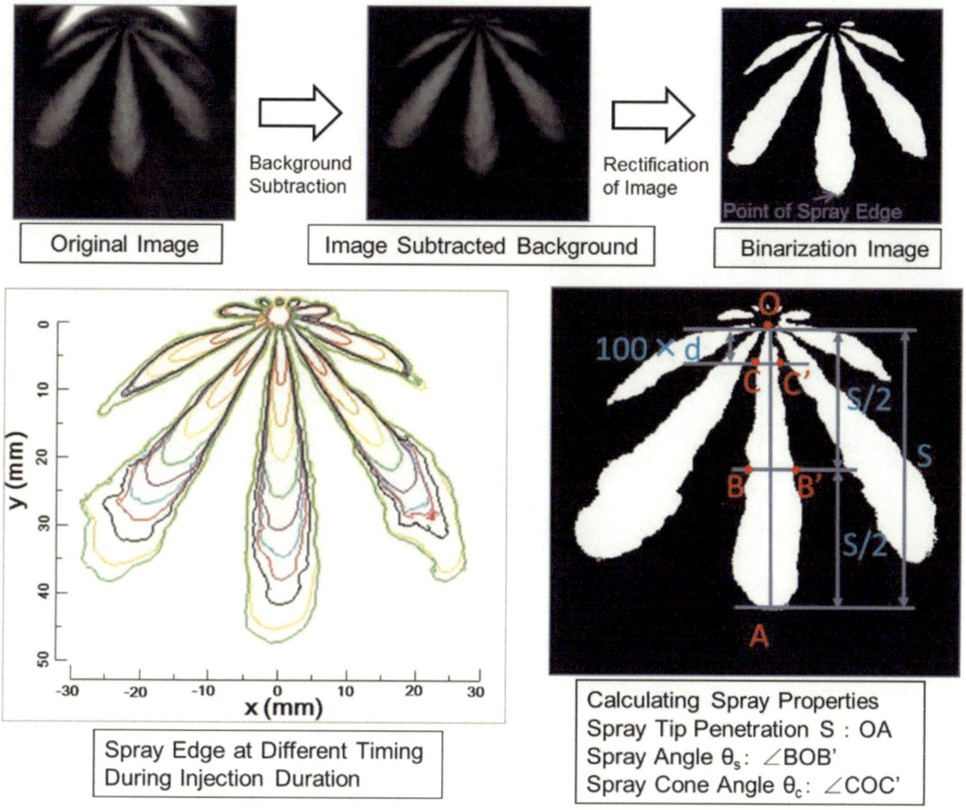

Figure 4.
Image processing for spray properties measurement.

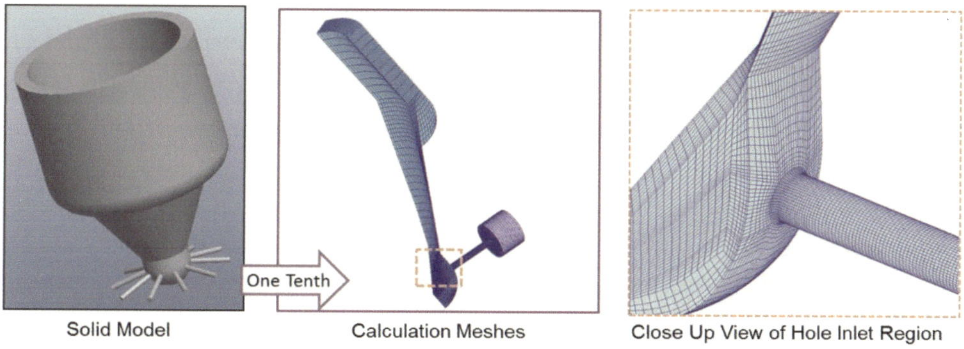

Figure 5.
Computational meshes.

sizes (7.769, 4.413, and 2.896 μm). During all the verification processes, it was proven that the distribution and occurrence of the cavitation could be predicted accurately relatively by this simulation.

Furthermore, aiming to make comparisons, the same transient needle-lift curve, as shown in **Figure 6**, measured from a similar type of multi-hole injector was applied to the two nozzle meshes.

Item		Classification	Setting/ Value
Model selection and Initial values	Turbulence model	*K-ζ-F*	
		Turbulence energy (m^2/s^2)	0.1
		Turbulence length scale (m)	0.000001
	Cavitation model	Linear cavitation model	
		Cavitation bubble density number	1.5 × 1018
		Diesel saturated vapor pressure (Pa)	892
Initial boundary	Inlet boundary	Injection Pressure (MPa)	120
	Export boundary	Ambient Pressure (MPa)	1.5
Mesh information	Needle-holder gap (mm)		0.002
	The minimum cell size in the hole region (mm)		4.413×10^{-3}
	The maximum grid number (cells)		132,796

Table 2.
Computational setting.

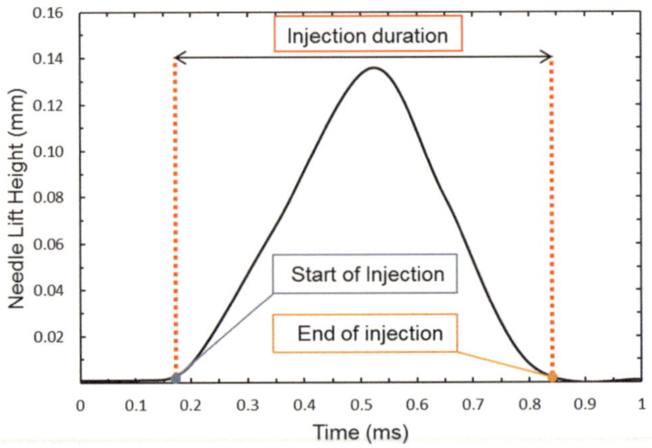

Figure 6.
Needle lift curve.

3. Results and discussion

3.1 Injection processes and spray characteristics

Figure 7 shows the injection rate results of the two different nozzles under the conditions of Q_{inj} = 2.0 mm^3/hole. These two curves present apparent distinctions. It can be seen that the micro-hole conspicuously changes the previous regularity of the fuel injection. The injection rate of the micro-hole (0.07 mm) nozzle is much lower than that of the nozzle with 0.10 mm hole diameter when maintaining the same injection quantity per hole. While the injection duration of the micro-hole nozzle is prolonged a lot. Moreover, the initial stage of the injection duration attracts attention

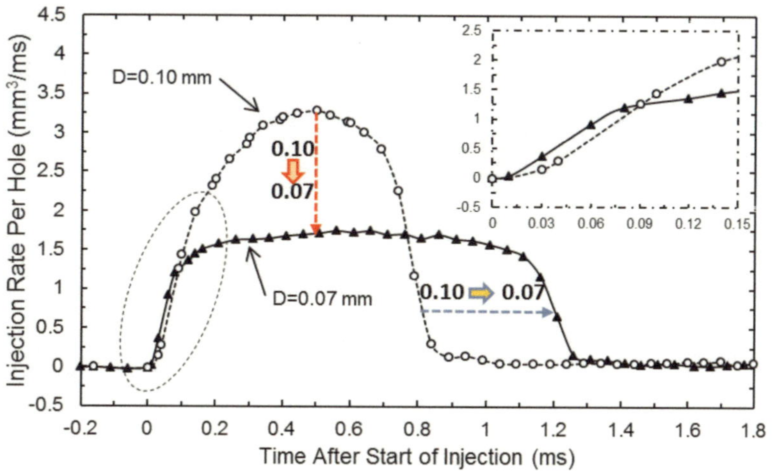

Figure 7.
The injection rate of different injectors.

to analysis in detail, which is emphasized and enlarged in **Figure 7**. The injection rate of the nozzle with a micro-orifice is a little higher than that of the other one with larger orifices in the initial stage of injection. In fact, it is known that the fuel injection velocity and the effective flow area alter the injection rate simultaneously. Furthermore, the fuel injection velocity mainly depends on the upstream pressure in the sac, and the effective flow area is affected by the hole numbers, hole diameter, and the discharge coefficient. The theoretical flow area of the micro-hole nozzle, whose diameter is 0.07 mm, is much smaller than that of the normal one (0.10 mm), and its sac pressure discharge rate should also be much lower than that of the nozzle with normal holes. As a result, all the phenomena described above reveal that the effect of micro-holes plays entirely different roles in the injection rate at different injection stages. Specifically, the higher sac pressure is mainly caused by the relatively lower sac pressure discharge rate at the initial stage of injection, and the injection rate of this stage is dominated by the consequent higher flow velocity inside the micro-holes. However, at the middle and post-stage injection, the relatively larger effective flow area and the consequent higher mass flow rate inside the nozzle with 0.10 mm hole diameter mainly dominate the injection rate of the nozzles.

Typical false-colored and temporal spray images of different nozzles are shown in **Figure 8**. According to the theoretical foundation of Mie scattering, the scattered light intensity is a symbolic characteristic of the droplet size and fuel concentration. The spray contours can help elucidate the effects of ambient gas entrainment and interactions between spray plumes [26].

As analyzed before, even the total injection mass per hole is held constant, the fuel injection quantity of the 0.10 mm hole is larger than that from the micro-hole at the same timing ASOI. Based on the Mie-scattering principle, the intensity is in proportion to the droplet size and concentration. Hence, the global intensity of the central sprays of the multi-hole nozzle with normal hole diameter, which are shown in **Figure 8(a)**, is much higher than that of the micro one, particularly at the beginning stage of fuel injection. The high-intensity area can even extend to the downstream region of the central spray. It follows that the fuel concentration of the micro-hole condition is leaner, and the atomization effect is better [27].

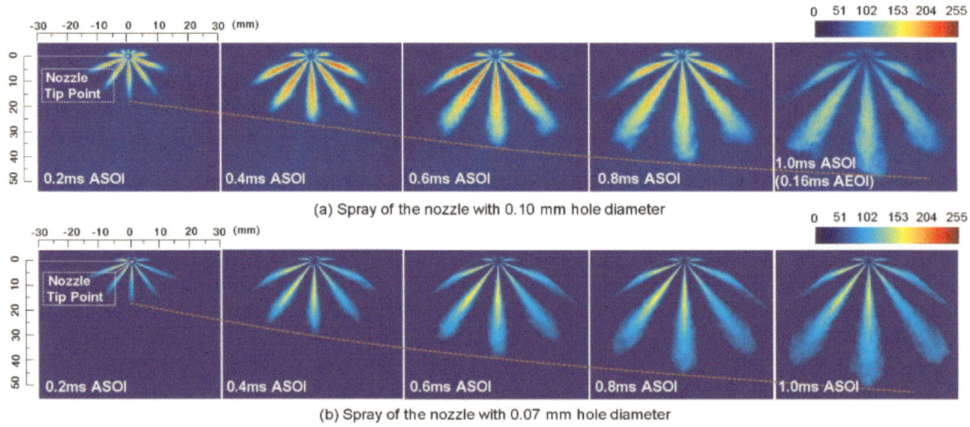

Figure 8.
False-colored spray images of different injectors.

As for the spray morphology, the edges of the spray of 0.10 mm holes are irregular compared with the micro-hole one, and the hole spray plumes are flanked by evident wavelike contours. In the case of micro holes, the edges of the spray upstream regions are neater and more orderly than those of nozzles with larger hole diameters.

Moreover, central sprays show a large eccentricity in the spray tip around the end of injection (0.8 and 1.0 ms ASOI) under the larger-hole nozzle condition, and the adjacent two plumes also present asymmetrical morphology. The low-pressure regime between the sprays generated by the air entrainment can enhance the sprawling diffusion of the multi-hole nozzle sprays. The Coanda effect [28] plays significant role in these phenomena. However, under the 0.07 mm hole diameter condition, the sprays have well symmetry. Therefore, it is safe to say that decreasing the hole diameter can supposedly reduce the uncontrollability and instability of the sprays emerging from multi-hole nozzles. A more specific investigation and discussion about this phenomenon will be introduced in the upcoming sections.

It is known that the spray propagation distance is governed by the upstream pressure and the ambient conditions. The calculated sac pressure and the measured spray tip penetration variation of different nozzles is shown in **Figure 9**. The Bernoulli equation is used to calculate the corresponding averaged sac pressure variations based on the injection rate measurement results in **Figure 7**. The equation is written as below, where Q_f is the injection rate, α is the average discharge coefficient, A represents the theoretical flow area, and Ps is the sac pressure.

$$Q_f = \alpha A \sqrt{2(P_s - P_a)/\rho_f} \tag{1}$$

As for α, it has much relationship with the local flow area, which is a direct reflection of the cavitation intensity of nozzle orifices. Nurick [29] and Payri et al. [30] conducted their experiments under quasi-steady conditions and concluded that the flow discharge coefficient of the nozzle hole mainly depended on its cavitation number under the cavitated conditions.

$$C_N = (P_{inj} - P_V)/(P_{inj} - P_a) \tag{2}$$

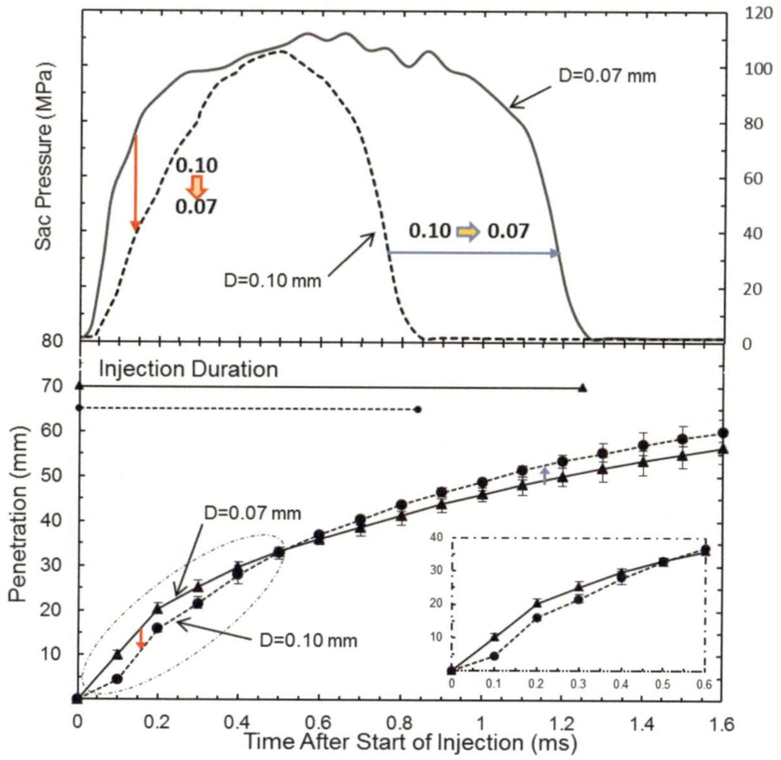

Figure 9.
Calculated sac pressure and spray pemetration variation of different injectors.

However, it is still difficult to get accurate quantitative results from the internal flow of a practical Diesel multi-hole nozzle, which is usually high pressure, high velocity, turbulent, and micro size. The theory developed under the quasi-steady condition was attempted to be expanded to the transient condition in the current study, and a comparative analysis was made between the different nozzles qualitatively to provide a reference for explaining different spray behaviors.

Throughout the injection duration, the pressure in the sac of the nozzle with micro-holes is all higher than that in the nozzle with normal holes. The different effective flow area, caused by the different hole diameter plays a significant role in this issue. As for the corresponding penetration result, it is in accordance with the injection rate variations analyzed previously. As the color arrows emphasize, the micro-hole nozzle spray tip penetration is longer at the initial stage. As the time elapses, the penetration of 0.10 mm hole nozzle passes over the micro-hole nozzle spray tip penetration. In different injection stages, it is the different factors (effective flow area or sac pressure) that dominate the spray propagation of nozzles with different holes. In the calibration processes of combustion system of Diesel engines, the fuel injection quantity, injection timing, and injection times/cycle are usually adjusted as the fuel supply strategy. The diffusion and deceleration of the multi-hole nozzle spray are usually associated with the transfer of the spray momentum to the turbulence energy [31]. Consequently, when concerning the optimization of the Diesel engine preference, effects of the micro-hole on fuel injection of multi-hole nozzles should be given attention emphatically.

Figure 10 shows the temporal variation of the spray angle and spray cone angle. It is evident that the spray angle is wider under the 0.1 mm hole condition. The maximum deviation appears at 0.1 ms ASOI, up to 11°. Furthermore, it seems that the micro-hole can exert more influence on the spray angle reduction compared with that on the spray tip penetration. On the other hand, different from the spray tip penetration results, no overlap happens among the two spray angle curves of nozzles with different holes. The simulation results in the upcoming section can be used to explain this phenomenon in depth.

The difference between the spray cone angles of the nozzles with different holes is also evident. It should be noted that the penetration of the larger-hole nozzle is shorter than 100 times the hole diameter at 0.1 ms ASOI; hence, only the spray angle can be measured, as shown in the figure. Because the position (10 mm) of 100 times of hole diameter is around the spray tip area of the 0.10 mm hole, it results in a small value of spray cone angle, and the spray cone angle difference between the two nozzles is not too much at 0.2 ms ASOI. After that, when the spray penetrates long enough, attributed to the more completed internal flow inside the larger holes [32], the 0.10 mm hole nozzle spray cone angle increases a great deal suddenly.

When it comes to the micro-hole condition, the value of the spray cone angle is much smaller and the variation is steadier. This can be explained like that since the cavitation collapse and turbulence flow inside different nozzles are two of the major mechanisms of the spray primary break-up [31, 33], the fuel jet enters the chamber

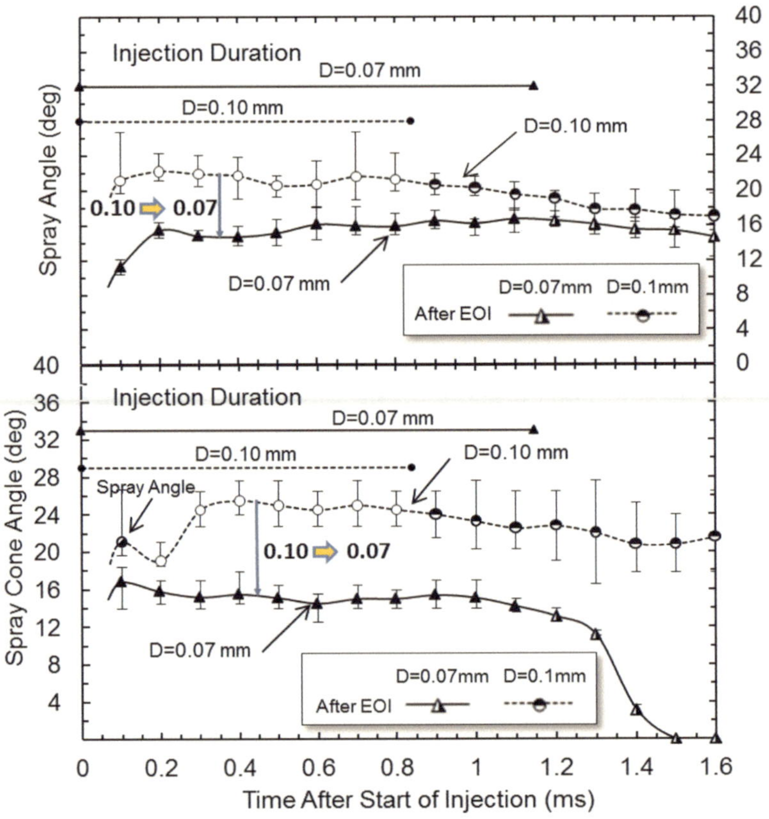

Figure 10.
Spray angle and cone angle variation of different injectors.

with a less cavitation level, reduced mass flow rate, momentum, and less turbulence caused by the increased ratio of nozzle hole length to diameter, which can result in the narrower spray cone angle.

According to the above discussion, because of the unique geometric structure, the micro-hole nozzle has a lower injection rate, higher sac pressure, and wider spray angle and spray cone angle compared to those of the nozzle with normal holes. Combing these phenomena with the interlaced relationship between the trends of the spray tip penetration of different nozzles, it is concluded that the effect of the micro-hole on different spray properties is discrepant.

3.2 Characteristics of the near-field spray

In order to investigate the spray behaviors near the nozzle tip region in detail, high-speed imaging of 100,000 fps was applied to take a close-up view of this regime. The Higher spatial and temporal resolutions allowed a more detailed observation of the very emergence of the fuel from the nozzle orifice. **Figure 11** shows the close-up gradient spray images of different nozzles, respectively.

According to the images of typical timings, the normal-hole nozzle sprays pulsate out from the nozzle tip to the radial direction, and the edges of them fluctuate seriously. The perturbation of the spray is marked and emphasized by different color arrows in the figure. On the other hand, since the injection duration is longer under the micro orifices condition when maintaining the same injection quantity, the selection for the typical timing of the images is a little different. With the same results under the imaging rate of 10,000 fps, the spray illumination intensity becomes weaker under the micro-hole condition. Of interest is that the spray pulsating phenomenon almost disappears, and the profiles of the sprays become much neater and more stable.

The angle, determined by 10 times the hole diameter away from the nozzle tip, is defined as the spray dispersion angle. The average spray dispersion angle and the single-shot results are shown in **Figure 12**. Generally, the spray dispersion angle under the hole diameter of 0.10 mm condition is much wider than that of the micro-hole one. As for the single-shot result, corresponding to the pulsating phenomenon of the near-field spray, as the capital letters and color arrows indicate in the figures, the spray dispersion angle curve waves and fluctuates strongly, especially in the initial stage of the injection (0.1–0.25 ms ASOI). The integrated speculations and reasonable explanations for this phenomenon can be excavated by linking the previous results [34] and the current study.

Different from the single-hole nozzle, due to the off-axis arrangement of the orifices, there is usually vortex flow inside the sac of multi-hole nozzles, especially under the condition of low needle lift. Moreover, with the needle moving, the location variation of the unstable vortex core results in the unstable spiraling flow pattern emerging through the hole with the vortex. During these processes, there is usually the generation of string-type cavitation in the sac and hole flow field. All the properties, only belonging to the internal flow of multi-hole nozzles, play significant roles in the phenomenon of near filed spray pulsating [7, 35]. When the hole diameter is reduced to 0.07 mm, the fluctuation of the spray dispersion angle curve decreases dramatically, and the angle becomes much narrower than that of the normal hole diameter condition. The reduced flow transverse can suppress the cavitation and vortex level under the micro-hole condition, and it is also impeded the fuel to enter into the micro-holes from the sac volume of the multi-hole nozzle. These could be

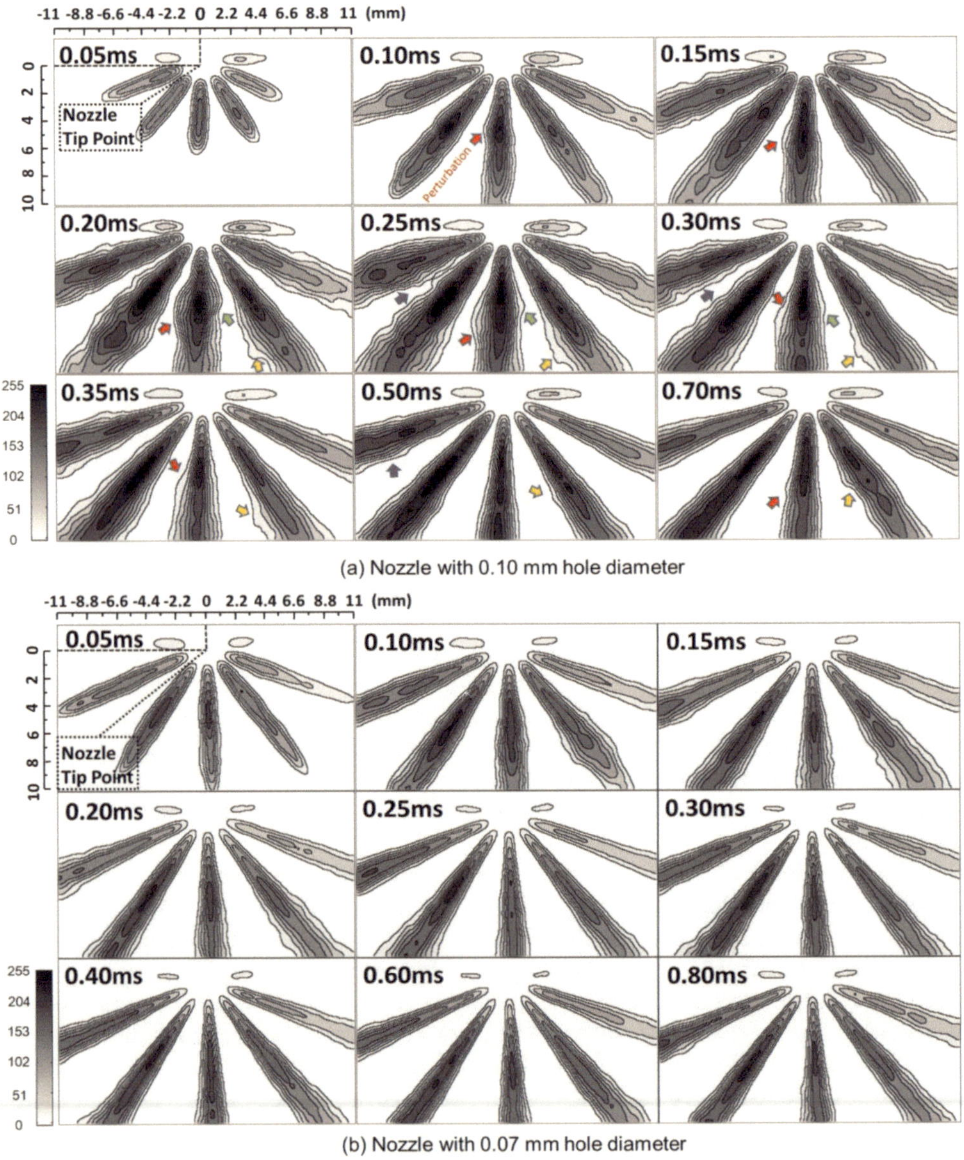

Figure 11.
Close-up view of sprays injected from different injectors.

used to explain the reason why the micro-hole multi-hole nozzle has a relatively narrower spray width and steady spray morphology.

3.3 Internal flow inside different nozzle configurations

The computational study is used to illustrate the different internal flow characteristics between the two nozzles. In this transient simulation analysis, according to the needle lift curve applied in the study, three typical timings (0.1, 0.318, and 0.5 ms ASOI) are selected to represent the initial, full needle lift, and post stages of the injection, respectively.

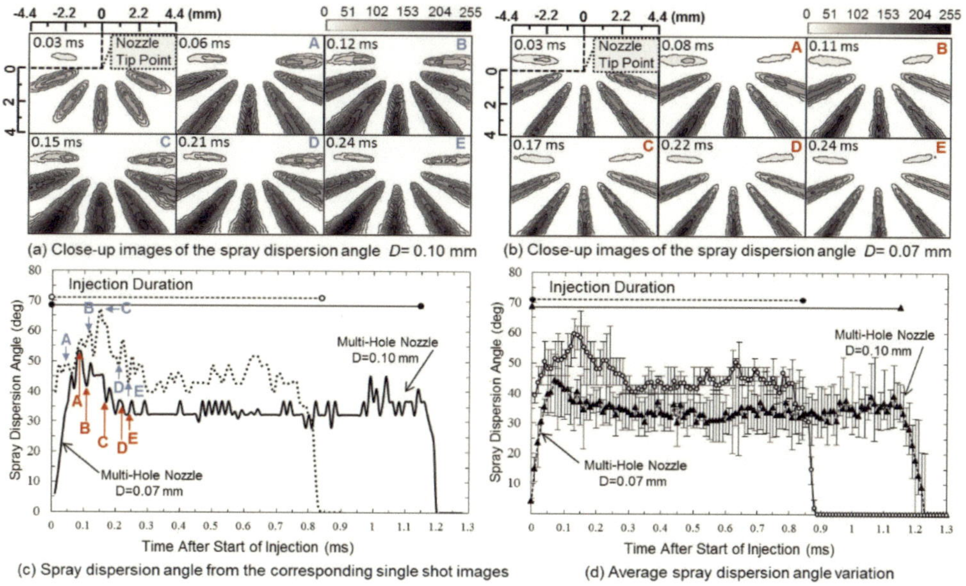

Figure 12.
Close-up spray behaviors injected from different injectors.

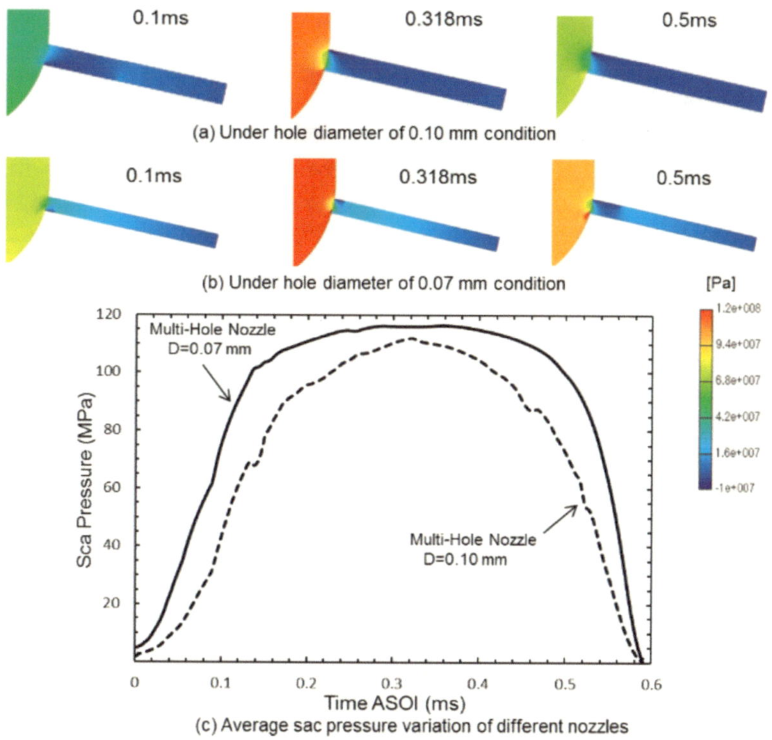

Figure 13.
Pressure distribution inside the nozzles at the typical timings during the injection duration.

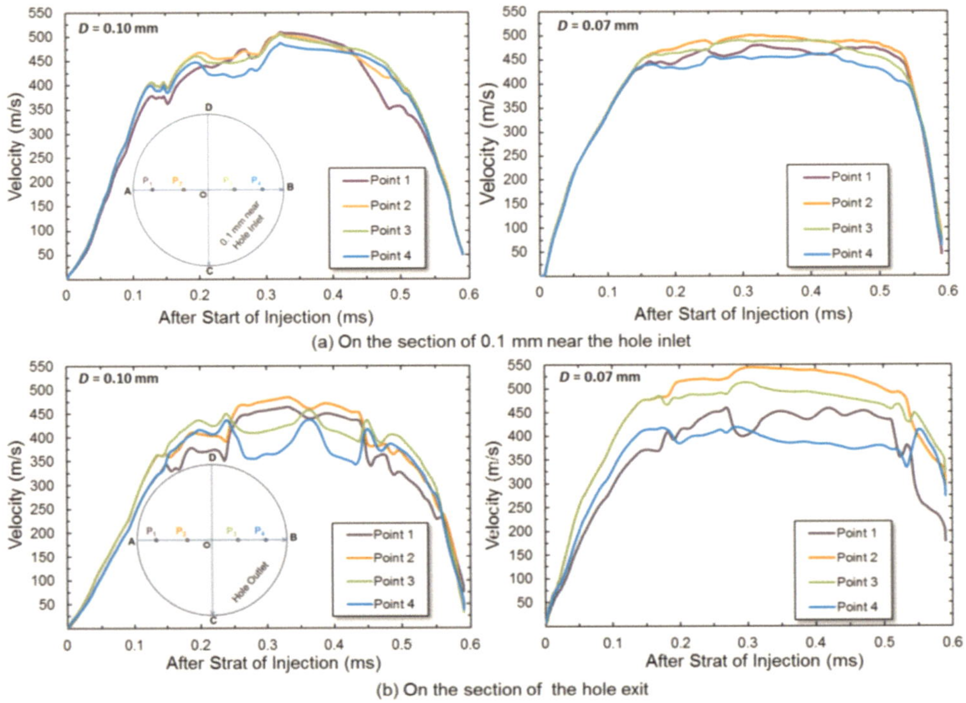

Figure 14.
Temporal variation of fuel velocity at the monitoring points on the hole sections.

Figure 13 shows the pressure variation inside different nozzles at the typical injection stages. The time-resolved averaged sac pressure variation of the two nozzles is also plotted in this figure. Attention should be paid to the nozzle with micro holes. Its sac and hole pressures are all higher than those of the one with larger orifices throughout the injection duration, which coincides with the experimental analysis shown in **Figure 9**. The lower theoretical effective flow area and sac pressure discharge rate of the micro-hole nozzle are mainly attributed to this issue.

The temporal velocity variation on the inlet and outlet sections of the two nozzles is shown in **Figure 14**. Four monitoring points (P1, P2, P3, and P4) were set along the horizontal diameter line (Line A-B) symmetrically.

The injection velocity on the outlet of the micro-hole nozzle is higher than that of the normal-hole one, especially in the initial stage of injection, which agrees with the discussion about the injection rate and sprays penetration results in the experiments. Furthermore, no matter on the inlet or on the outlet sections, the flow velocity fluctuation of the nozzle with larger orifices is the most intense one, which corresponds with the spray pulsating instance. Moreover, a reverse variation tendency appears at the symmetrical monitoring points with the needle moving up, which implies that there is a spiral and asymmetrical flow pattern inside the hole. However, under the micro-hole diameter conditions, the amplitude of the velocity wave decreases dramatically, which coincides with the neater spray profiles and the reduced spray pulsating phenomenon discussed in the experimental results.

The streamlines inside different holes, which are shown in **Figure 15**, can interpret the instance analyzed in the optical experiments. There are much more complicated streamlines with stronger curvatures and counter-rotating flow inside the hole volume

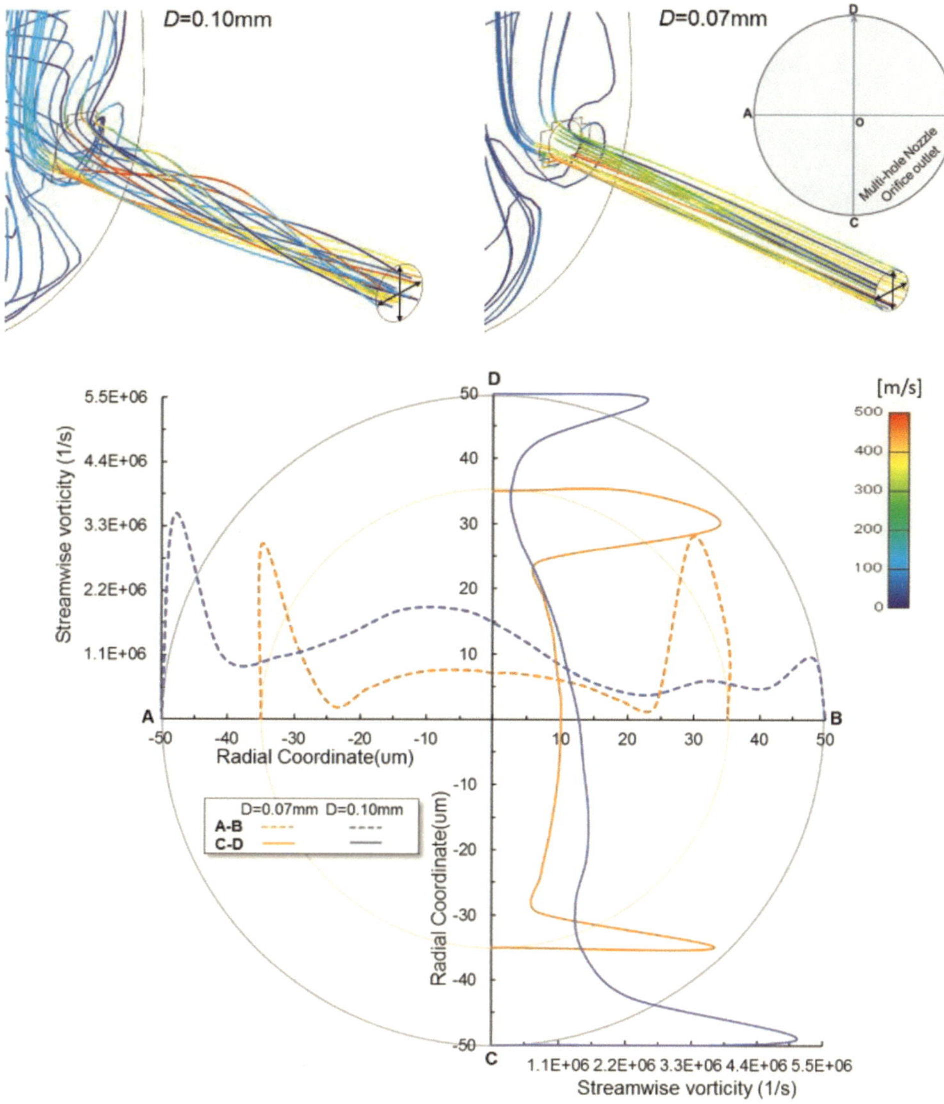

Figure 15.
Streamline and streamwise vorticity distribution along the horizontal and vertical diameter lines on the outlets of different multi-hole nozzles at full needle lift timing.

under the larger orifice condition. The vorticity distributions results along the orthogonal orifice lines also show higher stream-wise vorticity under 0.1 mm hole diameter conditions. This spiral flow is also observed by Gavaises and Andriotis [11], Lai et al. [32], and Hayashi et al. [35]. It has been proven that there are close correlations between this swirling motion and the wider spray cone angle. However, when attention is paid to the nozzles with micro-orifices, the vorticity decreases, and the streamline is stable and smooth relatively. This agrees with the reduced fluctuation of its spray behaviors observed in the experiments.

As shown in **Figure 16**, the velocity resolution of fuel jet injection on the hole exit section of the multi-hole nozzle is conducted, and V_a is the vector on the hole axis. The

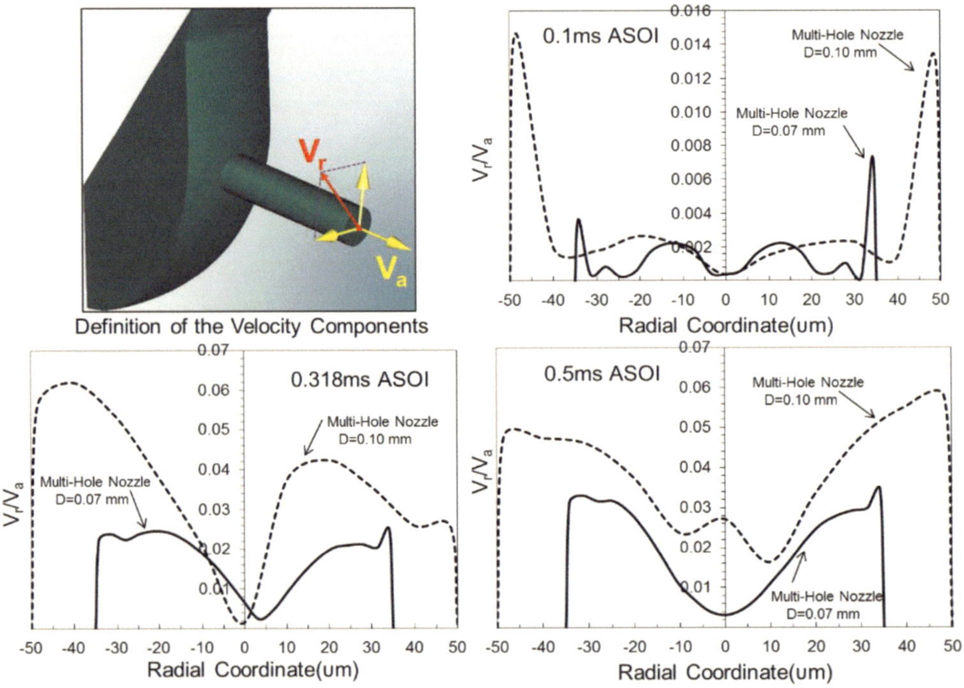

Figure 16.
Distribution of the ratio of velocity components (V_r/V_a) along the horizontal diameter of the exit of different nozzles at typical timings.

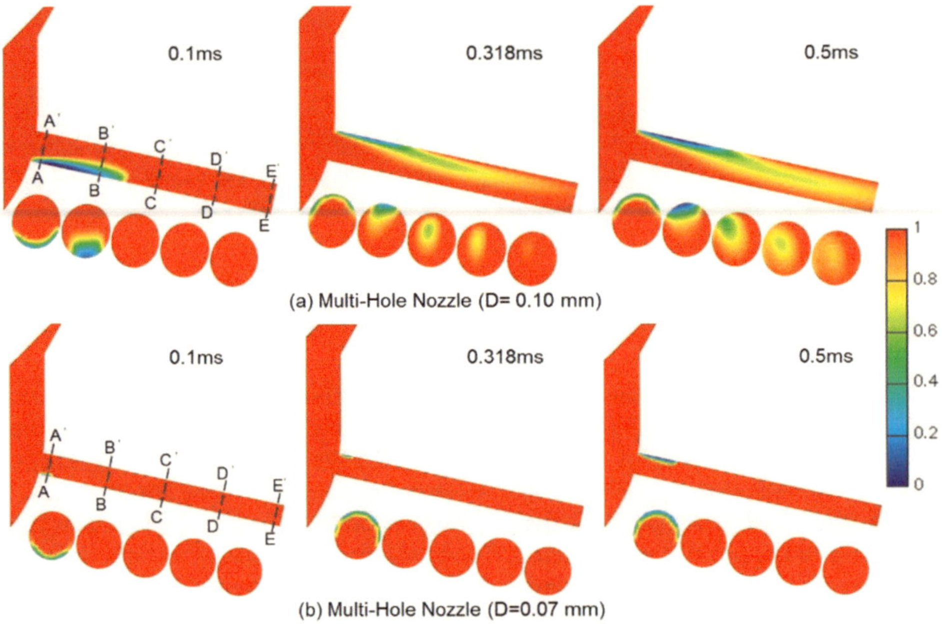

Figure 17.
Temporal variation of liquid volume fraction distribution inside different nozzle holes.

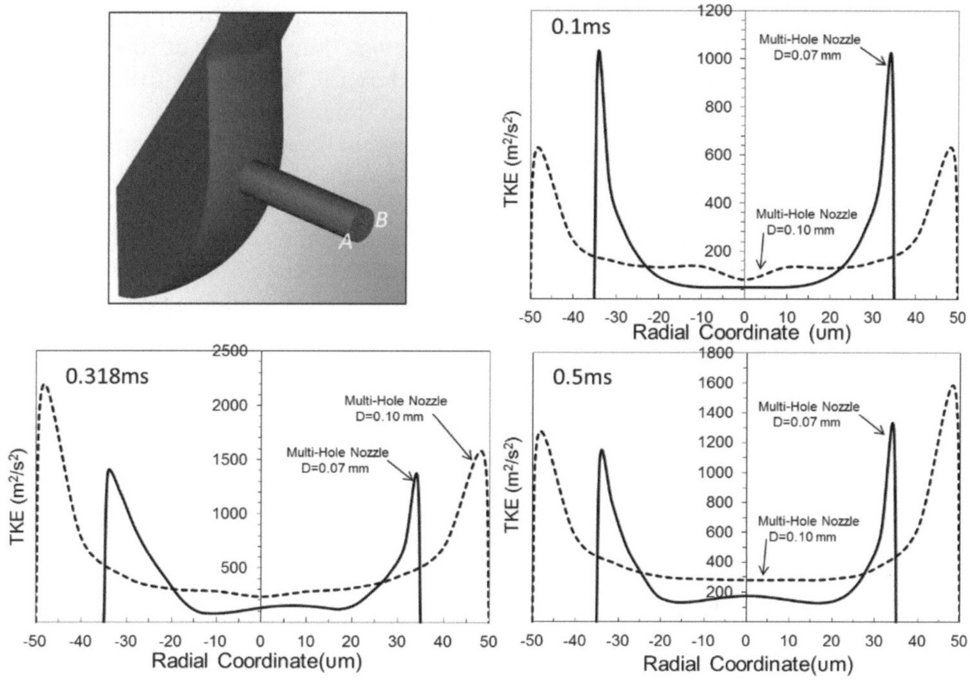

Figure 18.
Distribution of turbulence kinetic energy along the horizontal diameter of the exit of different nozzles at typical timings.

other velocity component V_r is on the outlet section, which could be divided into two components. In the current result, the ratio of V_r to V_a along the horizontal hole diameter line is shown in this figure. The normal-hole nozzle at three typical timings all has a much higher ratio than that of the micro-hole nozzle, which is coincided with the result observed in the experiments that the nozzle with 0.10 mm holes has a wider spray cone angle.

The temporal variations of the liquid volume fraction are shown in **Figure 17**. Affected by its special configuration, the cavitation distribution in the multi-hole nozzle is asymmetrical. Under the low need lift condition, when the injection starts, the film-type cavitation generates at a lower field of the hole inlet because of the aspects changing the flow direction. With the needle lifting up, the cavitation moves to the upper flow field and gradually develops into the string-type cavitation, which is mainly affected by the spiral and streamwise counter-rotating vortices flow structure inside the hole. Finally, the string-type cavitation can even reach the central area of the exit section, where intense mass and momentum transfer happens in the downstream region. The issues discussed above are all conducive to the wider spray cone angle [36]. However, under the micro-hole diameter condition, the cavitation intensity is much smaller, which is attributed to the higher pressure and smoother flow structure inside the hole volume of the nozzle.

The turbulence intensity increases with the enhancement of the dissipation of the spray momentum transferring to the turbulence energy, resulting in more intense liquid/gas interactions, lower spray velocity, and wider spray diffusion [37]. **Figure 18** shows the turbulence kinetic energy (TKE) distribution along the horizontal hole diameter line of different nozzles at the typical timing. The gradient of this value between the hole boundary locations and the hole central area of different nozzles is all prominent during the injection durations. However, they present different relations in the three injection stages. Initially, affected by the higher pressure increasing rate in the sac and the consequent higher velocity in the hole, the TKE value in the boundary location of the micro-hole is a little higher than that of the normal hole, but it is lower in the central area of the hole. After that, the TKE value under the normal-hole diameter condition increases a lot, and maintains a high level. However, the TKE value under the micro-hole diameter condition does not increase so much in the following injection stages, which indicates that the spray behaviors of the micro-hole nozzle would be more stable, resulting in the narrower spray angle and cone angle observed in the experiments.

According to the results of the internal flow of the different nozzles, except for the aerodynamic factors, it is the hydrodynamic factors and the unique properties of the internal flow in different nozzles that dominate their spray behaviors. The different characteristics of the flow patterns inside different nozzles mainly contribute to the deviation between their spray properties.

4. Conclusions

In the current work, the differences in the spray morphology between the two realistic multi-hole diesel nozzles under different hole diameter conditions were analyzed by the high-speed video observation method during the transient injection processes. The relationship between the internal flow, cavitation variation, and spray behaviors was investigated by the numerical simulations. The main conclusions are summarized as follows:

1. Different from normal single-hole nozzles, which are usually adopted in fundamental fuel injection research, unique and unstable spray behaviors of the multi-hole nozzles are observed in the experiments. The pulsating spray instance, and the wide spray angle and spray cone angle of multi-hole nozzles imply that the spray development can be affected greatly by its complex nozzle configurations. Sprays from the multi-hole nozzles are mainly dominated by the sac pressure, vortex flow in the sac, complicated spiral, and turbulent flow structure inside the hole except for the spray to spray interaction. The injection rate and spray tip penetration have a strong relationship with the pressure-increasing rate in the sac and the effective flow area of the nozzles.

2. The numerical simulation results about the internal flow of multi-hole nozzles with different hole diameters show that the increasing rate of sac pressure is faster inside micro-hole nozzles. Compared with the micro-hole, larger holes of the multi-hole nozzle can enhance the generation of dense swirling motion, where the string-type cavitation usually forms. Consequently, stronger vorticity, higher turbulence, and larger velocity components at hole exits are produced by these complex flow patterns, and the corresponding enhanced interfacial instability and wider spray propagation are observed in the optical experiments.

3. Influence of hole diameter on the internal flow, injection processes, and spray development of multi-hole nozzles is prominent. The reduced effective flow area suppresses the cavitation and turbulence flow, alters the injection rate, and

prolongs the injection duration. Moreover, the effect of hole diameter plays different roles in the spray properties (penetration, spray angle, and cone angle) of the multi-hole nozzles. The implications of these results have practical significance when considering the diesel fuel spray trajectory within the combustion chamber.

Acknowledgements

This research was supported by the Basic Research Operation Program of Dalian University of Technology (grant number 82232029).

Notes/thanks/other declarations

The authors would like to thank the Mazda Corporation for technical support and AVL-Japan for providing the numerical software.

Author details

PengBo Dong
School of Energy and Power Engineering, Dalian University of Technology, DaLian, China

*Address all correspondence to: pengbo.dong@dlut.edu.cn

References

[1] Bergwerk W. Flow pattern in diesel nozzle spray holes. Proceedings of the Institution of Mechanical Engineers. 1959;**173**:655-674. DOI: 10.1243/PIME_PROC_1959_173_054_02

[2] Hiroyasu H, Arai M, Shimizu M. Break-up length of a liquid jet and internal flow in a nozzle. In: ICLASS-91, 5th International Conference on Liquid Atomization and Spray Systems; July 1991; Gaithersburg, MD, U.S.A.: ILASS-Americas and National institute of standards and technology; 1991

[3] Koo J, Martin JK. Near-nozzle characteristics of a transient fuel spray. Atomization and Sprays. 1995;**5**:107-121. DOI: 10.1615/AtomizSpr.v5.i1.60

[4] Hiroyasu H, Arai M. Structures of fuel sprays in diesel engines. SAE Technical Paper. 1990:900475. DOI: 10.4271/900475

[5] Siebers DL. Liquid-phase fuel penetration in diesel sprays. SAE Technical Paper. 1998:980809. DOI: 10.4271/980809

[6] Araneo L, Coghe A, Brunello G, Cossali G. Experimental investigation of gas density effects on diesel spray penetration and entrainment. SAE Technical Paper. 1999;1999-01-0525. DOI: 10.4271/1999-01-0525

[7] Jin Y, Dong P, Zhai C, Nishida K, Ogata Y, Leng X. Internal flow and spray characterization of multi-hole injectors: Comparison with single-hole injectors. Energy & Fuels. 2020;**34**(6):7490-7501. DOI: 10.1021/acs.energyfuels.0c00473

[8] Arcoumanis C, Flora H, Gavaises M, Kampanis N, Horrocks R. Investigation of cavitation in a vertical multi-hole diesel injector. SAE Technical Paper. 1999;1999-01-0524. DOI: 10.4271/1999-01-0524

[9] Afzal H, Arcoumanis C, Gavaises M, Kampanis N. Internal flow in diesel injector nozzles: Modelling and experiments. In: Proceedings of the Institution of Mechanical Engineers Seminar on Fuel Injection Systems; 1–2 Dec 1999; London, United Kingdom: Institution of Mechanical Engineers; 1999. Paper S492/S2/99

[10] Arcoumanis C, Badami M, Flora H, Gavaises M. Cavitation in real-size multi-hole diesel injector nozzles. SAE Technical Paper. 2000;2000-01-1249. DOI: 10.4271/2000-01-1249

[11] Gavaises M, Andriotis A. Cavitation inside multi-hole injectors for large diesel engines and its effect on the near-field spray structure. SAE Technical Paper. 2006;2006-01-1114. DOI: 10.4271/2006-01-1114

[12] Dong P, Inaba T, Nishida K, Shimo D. Characteristics of the internal flow and the near-field spray of a single-hole injector and a multi-hole injector for diesel engines. Proceedings of the Institution of Mechanical Engineers, Part D: Journal of Automobile Engineering. 2016;**230**(5):632-649. DOI: 10.1177/0954407015591299

[13] Kastengren AL, Powell CF, Wang Y, Im K, Wang J. X-ray radiography measurements of diesel spray structure at engine-like ambient density. Atomization and Sprays. 2009;**19**:1031-1044. DOI: 10.1615/AtomizSpr.v19.i11.30

[14] Moon S, Gao Y, Park S, Wang J, Kurimoto N, Nishijima Y. Effect of the number and position of nozzle holes on in- and near-nozzle dynamic characteristics of diesel injection. Fuel.

2015;**150**:112-122. DOI: 10.1016/j.fuel.2015.01.097

[15] Schmide DP, Rutland CJ, Corradini ML, Roosen P, Genge O. Cavitation in asymmetric two-dimensional nozzles. SAE Technical Paper. 1999;1999-01-0518. DOI: 10.4271/1999-01-0518

[16] Masuda R, Fuyuto T, Nagaoka M, Von BE, Tatschl R. Validation of diesel fuel spray and mixture formation from nozzle internal flow calculation. SAE Technical Paper. 2005;2005-01-2098. DOI: 10.4271/2005-01-2098

[17] Andriotis A, Gavaises M, Arcoumanis C. Vortex flow and cavitation in diesel injector nozzles. Journal of Fluid Mechanics. 2008;**610**:195-215. DOI: 10.1017/S0022112008002668

[18] Som S, Ramirez AI, Longman DE, Aggarwal SK. Effect of nozzle orifice geometry on spray, combustion, and emission characteristics under diesel engine conditions. Fuel. 2011;**90**: 1267-1276. DOI: 10.1016/j.fuel.2010.10.048

[19] Bergstrand P, Denbratt I. Diesel combustion with reduced nozzle orifice diameter. SAE Technical Paper. 2001; 2001-01-2010. DOI: 10.4271/2001-01-2010

[20] Desantes JM, García-Oliver JM, Pastor JM, Ramírez-Hernández JG. Influence of nozzle geometry on ignition and combustion for high-speed direct injection diesel engines under cold start conditions. Fuel. 2011;**90**(11):3359-3368. DOI: 10.1016/j.fuel.2011.06.006

[21] Pickett LM, Siebers DL. Orifice diameter effects on diesel fuel jet flame structure. Journal of Engineering for Gas Turbines and Power. 2005;**127**(1): 187-196. DOI: 10.1115/1.1760525

[22] Dong P, Yamauchi J, Nishida K, Ogata Y. Hole geometrical effect on internal flow, fuel atomization and spray evaporation of multi-hole nozzle for diesel engine. SAE Technical Papers. 2017;2017-01-0860. DOI: 10.4271/2017-01-0860

[23] Bosch W. The fuel rate indicator: A new measuring instrument for display of the characteristics of individual injection. SAE Technical Paper. 1966; 660749. DOI: 10.4271/660749

[24] Mulemane A, Subramaniyam S, Lu PH, Han JS, Lai MC, Poola R. Comparing cavitation in diesel injectors based on different modeling approaches. SAE Technical Paper. 2004;2004-01-0027. DOI: 10.4271/2004-01-0027

[25] Blessing M, König G, Krüger C, Michels U, Schwarz V. Analysis of flow and cavitation phenomena in diesel injection nozzles and its effects on spray and mixture formation. SAE Technical Paper. 2003;2003-01-1358. DOI: 10.4271/2003-01-1358

[26] Dong P, Nishida K, Ogata Y. Characterization of multi-hole nozzle sprays and internal flow for different nozzle hole lengths in direct-injection diesel engines. Proceedings of the Institution of Mechanical Engineers, Part D: Journal of Automobile Engineering. 2017;**231**(4):500-515. DOI: 10.1177/0954407016653890

[27] Matsumoto Y, Gao J, Namba M, Nishida K. Mixture formation and combustion processes of multi-hole nozzle with micro orifices for D.I. diesel engines. SAE Technical Paper. 2007;2007-01-4049. DOI: 10.4271/2007-01-4049

[28] Skogsberg M, Dahlander P, Lindgren R, Denbratt I. Effects of injector parameters on mixture

formation for multi-hole nozzles in a spray-guided gasoline DI engine. SAE Technical Paper. 2005;2005-01-0097. DOI: 10.4271/2005-01-0097

[29] Nurick WH. Orifice cavitation and its effect on spray mixing. Journal of Fluids Engineering. 1976;**98**:681-687. DOI: 10.1115/1.3448452

[30] Payri F, Bermudez V, Payri R, Salvador FJ. The influence of cavitation on the internal flow and the spray characteristics in diesel injection nozzles. Fuel. 2004;**83**:419-431. DOI: 10.1016/j.fuel.2003.09.010

[31] Moon S, Gao Y, Wang J, Fezzaa K, Tsujimura T. Near-field dynamics of high-speed diesel sprays: Effects of orifice inlet geometry and injection pressure. Fuel. 2014;**133**:299-309. DOI: 10.1016/j.fuel.2014.05.024

[32] Lai MC, Zheng Y, Xie X, Moon S, Liu Z, Gao J, et al. Characterization of the near-field spray and internal flow of single-hole and multi-hole sac nozzles using phase contrast x-ray imaging and CFD. SAE Technical Paper. 2011;2011-01-0681. DOI: 10.4271/2011-01-0681

[33] Baumgarten C. Mixture Formation in Internal Combustion Engines. Berlin Heidelberg: Springer-Verlag; 2006. p. 45. DOI: 10.1007/3-540-30836-9

[34] Dong P, Nishida K, Inaba T, Ogata Y. Characterization of internal flow and spray Behaviors of hole-type nozzle under tiny and Normal injection quantity conditions for diesel engine. SAE International Journal of Fuels and Lubricants. 2016;**9**(1):125-137. DOI: 10.4271/2016-01-0862

[35] Hayashi T, Suzuki M, Ikemoto M. Effects of internal flow in a diesel nozzle on spray combustion. International Journal of Engine Research. 2013;**14**(6):646-654. DOI: 10.1177/1468087413494910

[36] Kubitschek JP, Weidman PD. Helical instability of a rotating liquid jet. Physics of Fluids. 2008;**20**:91-104. DOI: 10.1063/1.2973479

[37] Crua C, Heikal MR, Gold MR. Microscopic imaging of the initial stage of diesel spray formation. Fuel. 2015;**157**:140-150. DOI: 10.1016/j.fuel.2015.04.041